WESTERN FERTILIZER

HANDBOOK

Sixth Edition

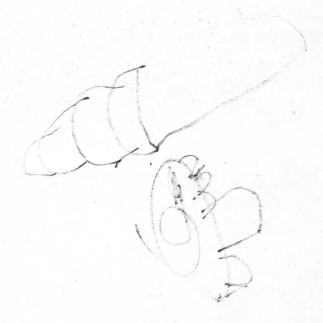

Western agriculture is dependent on water and fertilizer.

WESTERN FERTILIZER

HANDBOOK

Produced by:

SOIL IMPROVEMENT COMMITTEE
CALIFORNIA FERTILIZER ASSOCIATION

Order from:

THE INTERSTATE
PRINTERS & PUBLISHERS, INC.

Danville, Illinois 61832

Library of Congress Catalog Card No. 80-80127

ISBN 0-8134-2122-5

The sixth edition of the *Western Fertilizer Handbook* is dedicated to Earle J. Shaw, former chairman of the *Handbook* Committee. He, along with other leaders in agriculture, wrote, organized and published the first edition in 1953. Since then, more than 125,000 *Handbooks* have been placed into the hands of growers, students, advisors and professional agriculturists. Reprinted in both Spanish and French, the book has enjoyed worldwide recognition as a concise, accurate and authoritative treatise on irrigated agriculture. Although Earle has been "retired" for many years, his interest and efforts have not waned, as he has promoted the distribution of the *Handbook* to numerous agricultural schools and into various technical circles. Truly, Earle J. Shaw has been an influential spokesman for the *Handbook* throughout these many years.

TABLE OF CONTENTS

FOREWORD

Dramatic increases in national and worldwide demand for food, feed and natural fibers have brought increased pressures to bear on existing food and fiber production systems. Never before have agricultural exports been so important to our economy. They are a key factor in achieving a favorable balance of payments and stability of the dollar, in meeting the food needs of hungry nations and in providing a base for world peace.

The United States is the world's largest producer of agricultural products and the largest exporter of those products. Agriculture's continuing ability to produce will have an important effect on this nation's future political and economic life.

As the world population increases, we must produce more food, either by increasing the acreage devoted to agricultural crops or by increasing yields per acre. In the United States loss of agricultural land due to urbanization, the decreasing availability of water for irrigated agriculture, the increasing cost of energy utilized in agricultural production, and the increasing amount of regulation affecting agriculture, all have an impact on the food production capacity of this country. It seems clear that we must increase farm efficiency— productivity per acre—if we are to keep pace with the increasing claims on agriculture.

Plant nutrition and soil fertility are, of course, key factors in production capacity and efficiency, if wise decisions are made for their management. Accurate, current information is indispensable in making correct management decisions, and the California Fertilizer Association's Soil Improvement Committee is to be commended for providing it in this handbook.

As the primary source of technical information for growers and the fertilizer industry, the University of California's Division of Agricultural Sciences is pleased to participate in this endeavor. Expanding our food supplies depends on the availability of commercial fertilizers and on expanding knowledge of plant nutrition and plant life processes. As new knowledge is developed, the Division's scientists will continue to bring information to growers and the industrial agencies that serve them.

J. B. Kendrick, Jr.
Vice President—Agricultural Sciences
    and Director, Agricultural Experiment
    Station
University of California

ACKNOWLEDGEMENTS

Tremendous advances in agriculture technology have been made in recent years. Although most of the information in the previous editions of the *Western Fertilizer Handbook* is still current, much new information is now available. Thus, this sixth edition contains much of this information in the revised chapters and in a new chapter, Chapter 14. The writing and revising of this edition has been a team effort by many outstanding scientists. Those who have made major contributions include: Keith B. Campbell, W. A. Dickinson, Jr., George R. Hawkes, Einar Helgested, the late Malcolm H. Mc-Vickar, Wayne Morgan, H. M. Reisenauer, Paul Rogers, W. J. Sharratt, Earle J. Shaw, Robert Staib, L. K. Stromberg, J. T. Thorup and R. M. Thorup. Their contributions are greatly acknowledged.

The *Handbook* Committee has coordinated the excellent text submitted by the Association's contributors. Certainly, without their help this new edition would not have become a reality.

HANDBOOK COMMITTEE
Keith B. Campbell
Einar Helgested
Earle J. Shaw
George R. Hawkes, Chairman

INTRODUCTION

Man's primary need is for food; everything else plays a secondary or less important role. Man and animals depend either directly or indirectly upon plants for their food supply. The judicious use of fertilizer has a significant beneficial influence on plant growth. Sound usage of fertilizer increases crop yields and improves the quality of crops.

Sound fertilizer usage is a science in itself. The *Western Fertilizer Handbook* presents, in a straightforward manner and in concise language, information on soil, water, plant growth and on fertilizer products, their properties and how best to use them. Major emphasis is, however, placed on usage—usage as it applies to western soils.

WESTERN SOILS ARE DIFFERENT

Western agriculture is distinctly different in many ways from much of the agriculture east of the Rocky Mountains. Irrigation is widespread in the West. Because of the limited rainfall in most western areas, soils are not highly leached. Many of the soils have pH values in excess of 7.0. Most western soils are also well fortified with calcium, many are high in potash and a reasonable percentage are relatively high in phosphorus. Nitrogen and zinc deficiencies are relatively widespread in the West.

By contrast, most eastern agriculture is rain-fed. For the most part, eastern soils have been leached and, unless limed, have pH values under 7.0. The use of lime and potash along with nitrogen and phosphorus is a common practice in the East.

Many western farmers use amendments to aid in lowering pH values and/or to aid in leaching out undesirable substances. The chemical and physical characteristics of western soils greatly influence nutrient availability and fertilizer-soil relations.

This sixth edition of the *Western Fertilizer Handbook* presents information on the usage of soil amendments and management practices to adjust the soil's environment for greater fertilizer responses and favorable plant growth. Practices used for controlling or adjusting water quality have not been overlooked.

AGRICULTURE MORE THAN GROWING CROPS

Agriculture does not stop at the farm gate. The production of crops and livestock is only one segment of agriculture. At the production level, one farm worker produces enough food and fiber to meet the needs of more than 60 people. This efficiency is constantly increasing. Since this is an average farm figure, it seems safe to say that many highly efficient farmers—the type we have so many of in the West— produce enough food for at least 200 people.

The point to be made, however, is that many more people are employed in agriculture than just those on the farm. When one considers all aspects of agriculture, including processing, transportation, distribution and the services associated with agriculture, the picture changes drastically, and agriculture becomes our No. 1 industry, employing more people than any other industry. Unfortunately, many people actually employed in agriculture complain about food prices even though food prices, in relation to take-home pay here in the United States, are among the lowest in the world.

Telling the story of American agriculture, what it really encompasses and the fantastic job it does, is a responsibility of all people employed in agriculture.

Fortunately, recent developments have focused public attention on agriculture as a way of eliminating, or at least narrowing, our trade deficit. Now there is an appreciation, at high governmental levels, that this country has the capability of producing food and fiber in large amounts beyond its own needs. No other country can come close to matching the U.S. food production efficiency.

In brief, the reasoning goes like this: Why not trade agricultural commodities so badly needed by other countries for items such as crude oil which we need in large quantities? As a result of this new awakening, agriculture can and should take on a new and an improved image.

FERTILIZERS CONTRIBUTE GREATLY TO FOOD PRODUCTION

Fertilizers are credited with 30 to 40 percent of our total food production. The major proportion of this comes from nitrogen, the additional comes from phosphate, potash and other essential nutrients. Even more of our production will come from fertilizer in future years. Average acre yields in 1978 are compared to those of 1968 in the following table, illustrating the change during the last decade.

It seems safe to say that with time, more and more fertilizer will be necessary to produce food for our people and for exports. In addition, American people also take tremendous pride in their lawns and gar-

1968 Acre Yields vs. 1979 Yields

Crop	1968	1978	Percent Increase
Wheat, bu. (60 lbs.)	30.7	31.6	3
Rice, lbs.	4290	4518	5
Feed grains, tons	1.8	2.1	17
Peanuts, lbs.	1743	2630	51
Cotton, lbs. lint	436	520*	19
Hay, tons	2.1	2.3	10
Corn, bu. of grain (56 lbs.)	80	101	26
Soybeans, bu. (60 lbs.)	26.2	29	11
Grain sorghum, bu. (56 lbs.)	52.8	55	4

* 1977 yield.

dens. With more and more leisure time, additional emphasis will be placed on gardening, and this means more fertilizer usage for lawns and ornamentals. Fortunately, for mankind, as will be explained later, the judicious use of fertilizer also improves the environment.

CAREER OPPORTUNITIES ARE NUMEROUS

The career opportunities in agri-business, research, education and extension have never been brighter. The word *agribusiness* encompasses the activities associated with farming and the farm service industries.

Today, we see the real awakening as to the importance of efficient farm production and its value not only to people in the United States but as a factor in world trade. The demand for well-trained agriculturists will no doubt increase. Technical training is a must to qualify for employment in this, the most basic industry in the world. The primary fields are crop and animal production, farm operations, management, consultant or advisory service, soil sciences and agronomy, ornamental horticulture and the closely associated pesticide, fertilizer, seed, food and farm equipment industries.

Related fields of opportunity lie in crop processing, marketing, distribution, finance and agricultural journalism. For the more technically inclined student, many career opportunities lie in the fields of research, education and government, which are as satisfying and as remunerating as many other professions.

CALIFORNIA FERTILIZER ASSOCIATION DEDICATED TO SOUND AGRICULTURE

The California Fertilizer Association was organized in 1923 by a few farsighted leaders. The purposes of this organization are to promote the progress and development of the fertilizer industry in the interest of a sound agriculture, and more specifically, to produce and disseminate useful knowledge and information among its members and others interested in agriculture; or in the utilization of fertilizer, which would pertain to the scientific development of agriculture and to the increase of soil fertility; to encourage agriculture study and research in schools and colleges and efficiency in the commercial fertilizer industry; and to promote social and business intercourse among all persons interested in the improvement of soil and the production of fertilizers.

One of the first activities by the membership in the mid-1920's was the establishment of the Soil Improvement Committee. This committee was charged then, as now, with the responsibility of maintaining close liaison with official agriculture to further the proper use of fertilizer for the efficient production of quality food and fiber.

The Soil Improvement Committee produced the first edition of the *Western Fertilizer Handbook* in 1953. With later editions, total distribution has exceeded 125,000 copies. There are other agricultural chemical trade associations that sponsor promotional programs similar to those of the California Fertilizer Association. These include:

1. Arizona Agricultural Chemical Association, P.O. Box 505, Glendale, AZ 85311
2. Colorado Agricultural Chemical Association, P.O. Box 2171, Denver, CO 80201
3. Montana Plant Food Association, Department of Soils, Montana State University, Bozeman, MT 59715
4. National Agricultural Chemicals Association, 1155 - 15th Street, N.W., Washington, DC 20005
5. National Fertilizer Solutions Association, 8823 No. Industrial Road, Peoria, IL 61615
6. Northwest Plant Food Association, 1812 N.W. Kearney Street, Portland, OR 97209
7. Rocky Mountain Plant Food Association, P.O. Box 2171, Denver, CO 80201
8. The Fertilizer Institute, 1015 - 18th Street, N.W., Washington, DC 20036
9. The Sulphur Institute, 1725 K Street, N.W., Washington, DC 20006
10. Western Agricultural Chemical Association, 6650 Belleau Wood Lane, Suite 111, Sacramento, CA 95822

Chapter 1

SOIL—A MEDIUM FOR PLANT GROWTH

Food comes from the earth. The land with its waters
gives us nourishment. The earth rewards richly the knowing
and diligent but punishes inexorably in ignorant and sloth-
ful. This partnership of land and farmer is the rock foundation of our complex social structure.

W. C. Lowdermilk

WHAT IS SOIL?

As a medium for plant growth, soil can be described as a complex
natural material derived from disintegrated and decomposed rocks
and organic materials, which provides nutrients, moisture and an-
chorage for land plants.

The four principal components of soil are mineral materials, organic matter, water and air. These are combined in widely varying
amounts in different kinds of soil, and at different moisture levels. A
representative western soil, at an ideal moisture content for plant
growth, is nearly equally divided between solid materials and pore
space, on a volume basis. The pore space contains nearly equal
amounts of water and air. Figure 1-1 shows a schematic representa-
tion of such a soil.

HOW SOILS ARE FORMED

The development of soils from original rock materials is a long-
term process involving both physical and chemical weathering, along
with biological activity. The widely variable characteristics of soils
are due to differential influences of the soil formation factors:
1. Parent material—material from which soils were formed.
2. Climate—temperature and moisture.
3. Living organisms—microscopic and macroscopic plants and
   animals.
4. Topography—shape and position of land surfaces.
5. Time—period during which parent materials have been sub-
   jected to soil formation.

1

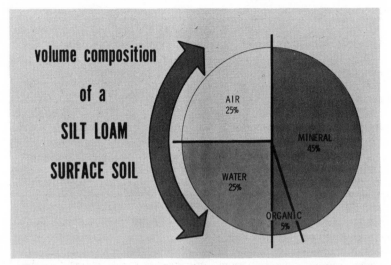

Fig. 1-1. Volumetric content of four principal soil components for a representative western soil at an ideal moisture content for plant growth.

The initial action on the parent rock is largely mechanical—cracking and chipping due to temperature changes. As the rock is broken, the total surface area exposed to the atmosphere increases. Chemical action of water, oxygen, carbon dioxide and various acids further reduces the size of rock fragments and changes the chemical composition of many of the resultant particulate materials. Finally, the action of microorganisms and higher plant and animal life contributes organic matter to the weathered rock material, and a true soil begins to form.

Since all of these soil-forming agents are in operation constantly, the process of soil formation is a continual one. Evidence indicates that the soils we depend on today to produce our crops required hundreds or thousands of years to develop. In this regard, we might consider soil as a nonrenewable resource, measured in terms of man's life span. Thus it is very important that we protect our soils from destructive erosive forces and nutrient depletion which can rapidly destroy the product of hundreds of years of nature's work.

SOIL PROFILE

A vertical section through a soil typically presents a layered pat-

tern. This section is called a "profile," and the individual layers are
referred to as "horizons." A representative soil has three general
horizons, which may or may not be clearly discernible (Figure 1-2).

The uppermost horizon includes the "surface soil" or "topsoil,"
and is designated the "A" horizon. The next successive horizon, underlying the surface soil, is called the "subsoil," or "B" horizon. The
combined A and B horizons are referred to as the "solum." Underlying the B horizon is the "parent material," or "C" horizon. These
three horizons, together with the unweathered, nonconsolidated rock
fragments lying on top of the bedrock, are called the "regolith."

Soil profiles vary greatly in depth or thickness, from a fraction of
an inch to many feet. Normally, however, a soil profile will extend to
a depth of about three to six feet. Other soil characteristics, such as
color, texture, structure and chemical nature also exhibit wide variations among the many soil types.

The surface soil (A horizon) is the layer which is most subject to
climatic and biological influences. Most of the organic matter accumulates in this layer, which usually gives it a darker color than the
underlying horizons. Commonly this layer is characterized by a loss
of soluble and colloidal materials, which are moved into the lower
horizons by infiltrating water, a process called "eluviation."

The subsoil (B horizon) is a layer which commonly accumulates
many of the materials leached and transported from the surface soil.
This accumulation is termed "illuviation." The deposition of such
materials as clay particles (colloidal material carried by infiltrating
water from the surface layer, as well as synthesized clay particles
formed from the recombining of soluble silicates and hydroxides),
iron, aluminum, calcium carbonate, calcium sulfate and other salts,
creates a layer which normally has more compact structure than the
surface soil. This often leads to restricted movement of moisture and
air within this layer, which produces an important effect upon plant
growth.

The parent material (C horizon) is the least affected by physical,
chemical and biological agents. It is very similar in chemical composition to the original material from which the A and B horizons were
formed. Parent material which has formed in its original position by
weathering of bedrock is termed "sedentary" or "residual," while
that which has been moved to a new location by natural forces is
called "transported." This latter type is further characterized on the
basis of the kind of natural force responsible for its transportation
and deposition. When water is the transporting agent, the parent
materials are referred to as "alluvial" (stream deposited), "marine"

Fig. 1-2.   This typical soil profile illustrates the three general soil horizons.

(sea deposited) or "lacustrine" (lake deposited). Wind-deposited materials are called "aeolian." Materials transported by glaciers are termed "glacial." And, finally, those that are moved by gravity are called "colluvial," a category that is relatively unimportant with respect to agricultural soils.

Because of the strong influence of climate on soil profile development, certain general characteristics of soils formed in areas of different climatic patterns can be described. For example, much of the western area has an arid climate, which results in the development of coarser textured soils (more sand particles) than most of those developed in more humid climates. Also, the soil profiles in many western soils are less developed, since the amount of water percolating through the soils is generally much less than in more humid climates. Because of this, many western soils contain more calcium, potash, phosphate and other nutrient elements than do the more extensively developed eastern soils.

Thus, the soil profile is an important consideration in terms of plant growth. The depth of the soil, its texture and structure and its chemical nature determine to a large extent the value of the soil as a medium for plant growth.

SOIL TEXTURE

Soils are composed of particles with an infinite variety of sizes and shapes. On the basis of their size, individual mineral particles are divided into three categories—sand, silt and clay. Such a division is very meaningful, not only in terms of a classification system but also in relation to plant growth. Many of the important chemical and physical reactions are associated with the surface of the particles. Surface area is enlarged greatly as particle size diminishes, which means that the smallest particles (clay) are the most important with respect to these reactions.

Two systems of classification of the various particle sizes (soil separates) are used. One is the U.S. Department of Agriculture system and the other is the International system. The size ranges of the various soil separates for the two systems are shown in Table 1-1.

Soil texture is determined by the relative proportions of sand, silt and clay found in the soil. Twelve basic soil textural classes are recognized. A classification chart based on the actual percentages of sand, silt and clay appears in Figure 1-3.

A textural class description of soils can tell a lot about soil-plant interactions, since the physical properties of soils are determined

Table 1-1. Size Limits of Soil Separates*

U.S. Department of Agriculture		International	
Name of Separate	Diameter (Range)	Fraction	Diameter (Range)
	(mm)		(mm)
Very coarse sand	2.0 —1.0		
Coarse sand	1.0 —0.5	I	2.0 —0.2
Medium sand	0.5 —0.25		
Fine sand	0.25—0.10	II	0.20—0.02
Very fine sand	0.10—0.05		
Silt	0.05—0.002	III	0.02—0.002
Clay	Below 0.002	IV	Below 0.002

*Source: USDA *Soil Survey Manual*, Agricultural Handbook No. 18.

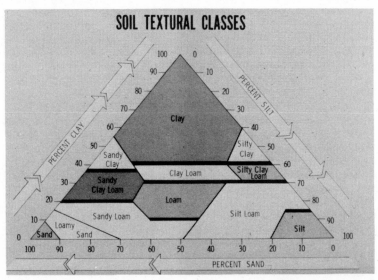

Fig. 1-3. Chart showing the percentages of clay (below 0.002 mm), silt (0.002 to 0.05 mm) and sand (0.05 to 2.0 mm) in the basic soil textural classes.

largely by the texture. In mineral soils, the exchange capacity (ability to hold plant nutrient elements) is related closely to the amount and kind of clay in the soil. The water-holding capacity is determined in large measure by the particle size distribution. Fine-textured soils (high percentage of silt and clay) hold more water than coarse-textured soils (sandy). Finer textured soils often are more compact, have slower movement of water and air and can be more difficult to work.

From the standpoint of plant growth, medium-textured soils, such as loams, sandy loams and silt loams, are probably the most ideal. Nevertheless, the relationships between soil textural class and soil productivity cannot be applied generally to all soils, since texture is only one of the many factors that influence crop production.

SOIL STRUCTURE

Except for sand, soil particles normally do not exist singularly in the soil, but rather are arranged into aggregates or groups of particles. The way in which particles are grouped together is termed "soil structure."

There are four primary types of structure, based upon shape and arrangement of the aggregates. Where the particles are arranged around a horizontal plane, the structure is called "plate-like" or "platy." This type of structure can occur in any part of the profile. Puddling or ponding of soils, often gives this type of structure on the soil surface. When particles are arranged around a vertical line, bounded by relatively flat vertical surfaces, the structure is referred to as "prism-like" (prismatic or columnar). Prism-like structure is usually found in subsoils, and is common in arid and semiarid regions. The third type of structure is referred to as "block-like" (angular blocky or subangular blocky), and is characterized by approximately equal lengths in all three dimensions. This arrangement of aggregates is also most common in subsoils, particularly those in humid regions. The fourth structural arrangement is called "spheroidal" (granular or crumb) and includes all rounded aggregates. Granular and crumb structures are characteristic of many surface soils, particularly where the organic matter content is high. Soil management practices can have an important influence on this type of structure.

Soil aggregates are formed by both physical forces and by binding agents—principally products of decomposition of organic matter. The latter types are more stable and resist to a greater degree the destructive forces of water and cultivation. Aggregates formed by physical forces such as drying, freezing and thawing, and tillage operations are relatively unstable and are subject to quicker decomposition.

Soil structure has an important influence on plant growth, primarily as it affects moisture relationships, aeration, heat transfer and mechanical impedance of root growth. For example, the importance of good seedbed preparation is related to moisture and heat transfer—both of which are important in seed germination. A fine granular structure is ideal in this respect.

The movement of moisture and air through the soil is dependent upon the porosity, which is influenced markedly by soil structure. Granular structure provides adequate porosity for good infiltration of water and air exchange between the soil and the atmosphere. This creates an ideal physical medium for plant growth. However, where surface crusting occurs, or subsurface claypans or hardpans exist, plant growth is hindered because of restricted porosity. Good management practices can improve soil structure and thereby create a better condition for crop production.

SOIL REACTION (pH)

The terms *acid, neutral* and *alkaline* refer to the relative concentrations of hydrogen ions (H^+) and hydroxyl ions (OH^-) in the soil solution. These concentrations are measured in terms of a pH value, which gives a measure of the active acidity in the soil solution, rather than the total or potential acidity of the soil. An acid soil has a higher concentration of hydrogen ions than hydroxyl ions, while an alkaline soil has the opposite. A neutral pH means that the two kinds of ions are present in equal amounts and counteract the effect of one another.

In order to distinguish between the relative degrees of acidity or alkalinity of a soil, a pH scale from 0 to 14 is used. At the middle of the scale (pH 7.0), soil is neutral in reaction, while below 7.0 the reaction is acidic, and above 7.0 it is alkaline (basic). The lower the pH value, the more acid the soil, and conversely, the higher the pH, the more alkaline. Since pH is a logarithmic function, each pH unit represents a tenfold increase or decrease in relative acidity or alkalinity. For example, a soil with a pH of 6.0 is 10 times more acid than one with a pH of 7.0. Also, a soil with a pH of 8.0 is 10 times more alkaline than one with a pH of 7.0 and 100 times more alkaline than one with a pH of 6.0.

The soil reaction is important to plant growth for several reasons:

(1) its effect on nutrient availability, (2) its effect on the solubility of toxic substances (such as aluminum), (3) its effect on soil micro-organisms and (4) the direct effect of pH on root cells (which affects the uptake of nutrients and water).

The relative availability of plant nutrient elements at various pH levels is shown in Figure 1-4. It is observed that a pH between 6.5 and 7.5 gives a maximum availability of the primary nutrients (N, P, K), and a relatively high degree availability of the other nutrient elements. For most agricultural crops, a soil pH of 6.0 to 7.0 is the most satisfactory range.

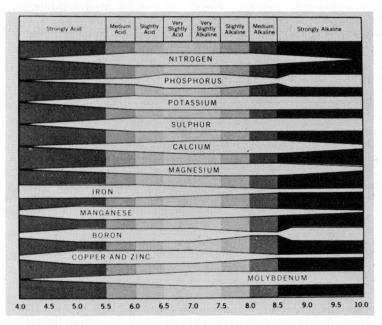

Fig. 1-4. How soil pH affects availability of plant nutrients.

Soils become more acid as a result of leaching the cations cal-cium, magnesium and potassium from the topsoil into the subsoil, and by removing these cations by growing crops. As the cations are removed from the soil particles, they are replaced with acid-forming hydrogen and aluminum ions.

Calcium carbonate in the soil acts as a buffer against acid formation, which means that it tends to restrict the development of acid soils. This occurs due to the increased solubility of calcium carbonate as the acidity increases, which increases the exchangeable calcium in the soil and removes hydrogen ions, which combine with oxygen from the carbonate to form water. Carbon dioxide is released in the process. For this reason, calcium carbonate, or agricultural limestone, is used as an amendment on acid soils.

Most common nitrogen fertilizers contribute to soil acidity, since their reactions in the soil increase the concentration of hydrogen ions in the soil solution. The increased use of fertilizers thus increases the acidulation process in soils and adds to the importance of soil testing to determine when corrective measures might become necessary.

CATION EXCHANGE CAPACITY

Due to their chemical structure, clay particles and decomposed soil organic matter (humus), are electrically active and carry a net negative charge. This means that electrically charged ions that carry a positive charge (cations) can be attracted to and held by these soil materials. Cations in the soil solution or adsorbed on the surface of plant roots can exchange positions with those adsorbed on the surface of clays or humus materials. The cation exchange capacity of a soil is a measure of the quantity of such cations that can be adsorbed or held by a soil.

Since soils contain varying amounts of and different kinds of clay and humus the total exchange capacity is widely variable among soil types. Most western soils contain predominantly the montmorillonitic and hydrous mica type clays, which have an exchange capacity approximately 5 to 10 times as great as that of the kaolinitic clays typical of soils in the southeastern United States. Because of the high exchange capacity of soil humus materials, soils with a high percentage of organic matter typically have higher exchange capacities than those of low organic matter content with similar amounts and types of clay.

The cations of greatest significance with respect to plant growth are calcium (Ca^{++}), magnesium (Mg^{++}), potassium (K^+), ammonium (NH_4^+), sodium (Na^+) and hydrogen (H^+). The first four are plant nutrients and are, therefore, involved directly with plant growth. The latter two have a pronounced effect upon nutrient and moisture availability.

The relative amount of each of the cations adsorbed on clay particles is closely associated with important soil properties. Highly acid soils have a high percentage of adsorbed hydrogen, while soils with a favorable pH (6.0 to 8.0) have predominantly calcium ions. Soils that are high in sodium ions are dispersed and resist the infiltration of water, while those with a high percentage of calcium ions are flocculated and favor higher infiltration rates. Mineral soils with a high exchange capacity are typically more fertile than those with lower exchange capacities, since they resist the loss of plant nutrients through the leaching process.

SOIL MICROORGANISMS

Besides their role in soil-forming processes, soil organisms make an important contribution to plant growth through their effect on the fertility level of the soil. Particularly important in this respect are the microscopic plants (microflora) which function in decomposing organic residues and releasing available nutrients for growing plants.

Some important kinds of microorganisms are bacteria, fungi, actinomycetes and algae. All of these are present in the soil in very large numbers, when conditions are favorable. A gram of soil (about one cubic centimeter) may contain as many as 4 billion bacteria, 1 million fungi, 20 million actinomycetes and 300,000 algae. These microorganisms are important in the decomposition of organic materials, the subsequent release of nutrient elements and the fixation of nitrogen from the atmosphere.

Soil bacteria are of special interest because of their many varied activities. In addition to the group of bacteria which function in decomposing organic materials (heterotrophic bacteria), there is a smaller group (autotrophic bacteria) which obtain their energy from the oxidation of mineral materials such as ammonium, sulfur and iron. This latter group is responsible for the nitrification process (oxidation of ammonium to nitrate nitrogen) in the soil, a process which is vitally important in providing nitrogen for the growth of agricultural crops.

Nitrogen-fixing bacteria also play an important role in the growth of higher plants, since they are capable of converting atmospheric nitrogen into useful forms in the soil. Nodule bacteria (rhizobia) live in conjunction with roots of leguminous plants, deriving their energy from the carbohydrates of the host plants, and fix nitrogen from the soil atmosphere. Free-living bacteria (azotobacter and clostridium) also fix atmospheric nitrogen, although to a lesser extent than the rhizobia bacteria, under most conditions.

Because of the important contributions made by bacteria to the fertility level of the soil, it has been stated that if their functions were to fail, life for higher plants and animals would cease.

SOIL MANAGEMENT

"To use to the best advantage" is the definition of the word *manage* which best applies to soil management. This implies using the best available knowledge, techniques, materials and equipment in crop production. Through wise management, the farmer can produce crops in the abundance required to feed the growing population, and at the same time improve soils and the environment, thus providing a priceless legacy for future generations.

Tillage is one of the important management practices used in agriculture. It serves many purposes—including seedbed preparation,

Fig. 1-5. Tillage is one of the most important management practices used in agriculture.

weed control, incorporation of crop residues and fertilizer materials, breaking soil crusts and hardpans to improve water penetration and aeration and shaping the soil for irrigation and erosion control. Because of the potential damage to soil structure from overworking the soil and for economic and fuel conservation purposes, the modern approach is to use only as much tillage as is required to produce a good crop. The term *minimum tillage* is applied to this concept. It should be emphasized, however, that the minimum is determined by the type of crop, the soil type and field conditions. No one set of guiding standards is appropriate for all situations.

Soil conservation is another important management practice which deserves close attention. It is estimated that annually in the U.S. 4 billion tons of sediment are lost from the land in runoff waters. That is equivalent to the total loss of topsoil (6 inches deep) from 4 million acres. Wind erosion is also a problem in certain areas, particularly in arid regions. Management practices such as contouring, strip planting, cover-cropping, reduced tillage, terracing and crop residue management help to eliminate or minimize the loss of soil from water and wind erosion. In addition to these practices, a sound fertilizer program promotes optimal growth of crops, which contributes to soil erosion control by protecting the soil against the impact of falling rain and holding the soil in place with extensive plant root systems.

Proper utilization of crop residues can be a key management practice. Crop residues returned to the soil improve soil productivity through the addition of organic matter and plant nutrients. The organic matter also contributes to an improved physical condition of the soil, which increases water infiltration and storage and aids aeration. This is vital to crop growth, and it improves soil tilth. In deciding how to best utilize crop residues, the immediate benefits of burning or removal should be weighed against the longer-term benefits of soil improvement brought about by incorporation of residues into the soil.

Special consideration should be given to the environmental aspects of soil management. Comments regarding soil erosion in terms of soil losses have already been made. The environmental implications of erosion are extremely important, since sediment is by far the greatest contributor to water pollution. Management practices which minimize soil erosion losses, therefore, contribute to cleaner water. The judicious use of fertilizers, which includes using the most suitable analyses and rates of plant nutrients, as well as the proper timing of application and placement in the soil, is also important.

Only when improperly used are fertilizers a potential pollution hazard. When used judiciously, they can make a significant contribution to a cleaner, more enjoyable environment. These effects are noted in three areas—cleaner air, cleaner water and improved wildlife habitat.

SUPPLEMENTARY READING

1. *Fundamentals of Soil Science*, Fifth Edition. H. D. Foth and L. M. Turk. John Wiley & Sons, Inc. 1972.
2. *The Nature and Properties of Soils*, Eighth Edition. N. C. Brady. The Macmillan Company. 1974.
3. *Soil Genesis and Classification*. S. W. Buol, F. D. Hole and R. J. McCracken. Iowa State University Press. 1973.
4. "Soil Physical Conditions and Plant Growth." *Agronomy*, Vol. II. Academic Press, Inc. 1952.
5. *Soils, an Introduction to Soils and Plant Growth*, Third Edition, R. L. Donahue, J. C. Shickluna and L. S. Robertson. Prentice-Hall, Inc. 1971.
6. *Soil Survey Manual.* USDA Agricultural Handbook No. 18. Soil Survey Staff. U.S. Government Printing Office. 1951.

Chapter 2

WATER AND PLANT GROWTH

All water used for irrigation contains some dissolved salts. The suitability of water for irrigation generally depends on the kinds and amounts of salts present. All salts in irrigation waters have an effect on plant-soil-water relations, on the properties of soils and indirectly on the production of plants.

A user of irrigation water should know the effects that water quality and irrigation practices have on:

1. The salt content (salinity) of the soil.
2. The sodium status of the soil.
3. The rate of water penetration into the soil.
4. The presence of elements which may be toxic to the crops.

It is difficult to isolate each of these factors from one another because some of them are interrelated.

IRRIGATION

Irrigation water is applied to soil to replenish the water removed from the soil by evaporation, by growing plants and to a lesser extent by drainage below the root zone. It is applied in a number of ways. The method of application chosen depends upon the crop to be grown, the depth and texture of the soil, the topography of the land, the cost of water and a number of other economic and physical reasons. The amount of water used and how often it is used is determined by the needs of the crop and the need to provide deep leaching occasionally to prevent accumulation of salts within the root zone. Therefore, successful irrigation requires careful management of both crops and water.

SOIL MOISTURE BEHAVIOR

In a well-drained soil, water is held largely as a film around each soil particle. The thinner these films are, the more tightly the water is held and the greater the suction needed to remove the water.

Immediately following an irrigation, the films of water are thick

15

Fig. 2-1. Irrigation water is applied to replenish the water removed from the soil. Photo by Valmont Industries, Inc.

and, therefore, are not tightly held on the soil (there is less suction needed to pull the water from these films).

In about two or three days, with free drainage, about one-half of this weakly held water moves deeper into the soil, and free drainage practically ceases. The moisture content at this point is called the *field moisture capacity*. The films of water are now thinner and are held more tightly.

Below the field moisture capacity, gravity is no longer a significant force in moving water in the soil; it is replaced by the roots of growing plants. Plants will remove about one-half of the water held at the field moisture capacity. At this point the films of water are only one-half as thick, and the soil now holds moisture so tightly that plants cannot extract it, thus causing them to wilt. This point is called *permanent wilting percentage*.

The moisture content of a soil saturated with water, its *saturation percentage*, is about twice the field moisture capacity and about four times the permanent wilting percentage. This relationship between the saturation percentage, field moisture capacity and permanent wilting percentage is accurate for practical purposes for all soils from clay loams to sandy loams.

A very moist soil has a thick film of water and hence has low suction. A drier soil has a thin film of water and has a high suction. For this reason, water will move from a wet soil to a drier soil, but the rate of such movement is slow.

WHEN TO IRRIGATE

Plant appearance is often used as a guide to follow in determining when to irrigate. Symptoms such as slowing down of growth, "bluish" color of leaves and temporary afternoon wilting are signs of moisture stress. Usually irrigation should be done before these signs dominate.

A soil tube or an auger can show depth of wetting and depletion of moisture. Soil moisture content may be measured by placing weighed samples of moist soils in an oven to determine the percent of moisture loss. Experienced growers can determine fairly accurately the moisture available by feeling soil samples taken with a tube or an auger.

Moisture measuring devices such as tensiometers and electrical resistance blocks indirectly measure soil moisture and are used extensively in research work. Properly placed, they can be useful in determining the moisture levels in the soil and scheduling irrigations.

A *tensiometer* consists of a porous ceramic cup imbedded in the major root zone of the soil. A rigid tube is connected to the cup and to a vacuum gauge just above the soil surface. The whole system is filled with water and sealed tightly. As the roots remove water from the soil, the soil draws water from the porous cup, creating a suction which is measured by the gauge. The drier the soil, the greater the suction. Tensiometers are effective for suctions from 0 to 0.8 of an atmosphere (80 centibars on most gauges).

Electrical resistance blocks consist of one or more pairs of electrodes imbedded in a porous material, usually gypsum. The blocks are buried in the root zone with wires leading to the soil surface. As the soil moisture changes, the moisture in the gypsum block changes with it. When the soil is wet, the electrical resistance is low. As the soil and block become dry the resistance increases.

This resistance is read by a specially designed meter attached to the wires. The meter is small and easily portable and can be used to monitor many stations. Gypsum blocks are usually not effective at soil tensions less than 2 atmospheres. Soluble salts in the soil solution decrease the electrical resistance and indicate a higher moisture content than is actually present.

Although tensiometers and resistance blocks are useful, they are limited to moisture readings in predetermined locations and depths only. These readings need to be made at regular frequent intervals.

The number of instruments needed, the depth of placement and the interpretation of the readings depend upon the soil variability and upon the crop and irrigation system. Measurements of soil moisture levels provide only indirect measurements of plant moisture stress. Recently, some sensors have been developed to measure directly plant moisture stress in the field. These sensors can be used with confidence on some plants to determine the optimum time of irrigation. Hopefully, future research will refine the techniques to make these measurements more widespread.

WATER ANALYSIS TERMINOLOGY

Various terms and units are used in reporting a chemical analysis of water. An understanding of the commonly used methods of reporting is needed to interpret the data properly.

Dissolved salts are dissociated into electrically charged particles called *ions*. The ions which are positively charged are called *cations* while the negatively charged ions are called *anions*. The concentration of each of these ions in the water is the usual method of reporting.

The common cations reported by laboratories are: calcium (Ca), magnesium (Mg), sodium (Na) and potassium (K). The anions usually reported are: bicarbonate (HCO_3), carbonate (CO_3), chloride (Cl) and sulfate (SO_4).

Substances usually found in smaller amounts are: boron (B) and nitrate (NO_3), which is sometimes reported as nitrate-nitrogen (NO_3–N).

Three principal methods are used to express the concentration of constituents in water. They are grains per gallon, parts per million and milliequivalents per liter.

Grains per gallon uses the English system of units. Most laboratories no longer use this method of reporting for irrigation water. It is still used to report hardness of water. To change grains per gallon to parts per million, multiply grains per gallon by 17.2.

Parts per million (ppm) is defined as 1 part of a salt to 1 million parts of water—or 1 milligram of salt per kilogram of solution. Water analyses are usually reported on a volume basis. One liter

(1.057 quart) of water weighs 1 kilogram (1,000 grams or 2.2 pounds); therefore, 1 milligram (one-thousandth of a gram) per liter is 1 part per million. Parts per million is used to report boron, nitrates, nitrate-nitrogen and a few other materials found in relatively small amounts.

Milliequivalents per liter (me/1) is the most meaningful method of reporting the major constituents of water. This is a measure of the chemical equivalence of an ion.

Salts are combinations of cations (sodium, calcium, magnesium, etc.) and anions (chlorides, sulfates, bicarbonates, etc.) in definite weight ratios. These weight ratios are based upon the atomic weight of each constituent and upon the valence (electrical charge). An equivalent weight of an ion is its atomic weight divided by its valence. A milliequivalent is 1/1,000th of an equivalent. (See Table 2-1 for the approximate equivalent weights of the common ions.)

Some laboratories report the common constituents in parts per million. To convert to milliequivalents per liter, divide the ppm by the equivalent weight of an ion. For example, a laboratory reports 54 ppm of sulfate. The equivalent weight of sulfate found in Table 2-1 is 48. Divide 54 by 48 and find the water has 1.12 me/1 of sulfate.

Table 2-1.   Major Constituents in Irrigation Water

	Ion Name	Symbol	Equivalent Weight
			(gm)
Cations	Calcium	Ca^{++}	20
	Magnesium	Mg^{++}	12
	Sodium	Na^{+}	23
	Potassium	K^{+}	39
Anions	Bicarbonate	HCO_3^{-}	61
	Carbonate	$CO_3^{=}$	30
	Chloride	Cl^{-}	35.5
	Sulfate	$SO_4^{=}$	48

The *pH* expresses the acidity or alkalinity of water. A pH reading of less than 7.0 is acidic, 7.0 is neutral and above 7.0 is on the alkaline side. Most well waters range from pH 7.0 to pH 8.5. Some stream waters may be as acidic as pH 6.5. The pH measurement is not important if the major ions are reported and is often omitted from laboratory reports.

Total salt content is usually reported as the electrical conductivity (EC). Chemically pure water does not conduct electricity, but water with salts dissolved in it does. The more salt in the water, the better conductor it becomes. The ability of a water sample to conduct electricity is used to determine its salt content. EC is usually reported as millisiemens per centimeter (ms/cm).*

The total salt content or total dissolved solids (TDS) is usually reported as ppm. This can be approximated by evaporating a known weight of water sample to dryness and weighing the salt remaining. More often it is approximated by measuring the EC in ms/cm and multiplying by 640. This gives the total dissolved solids in ppm. If a water has an EC of 1.6, then $1.6 \times 640 = 1,024$ ppm total dissolved solids. The me/l of total salts can be closely estimated by multiplying the EC in ms/cm by 10. For example, a water with an EC of 2.62 ms/cm contains 26.2 me/l of total salts.

Electrical conductivity is commonly used to check the salt content of soils. The conductivity is measured on the saturation extract from the soil and is designated EC_e. This is used to monitor the changes in the salt content of the soil resulting from irrigation. It is also useful in evaluating the relative tolerance of plants to salt and the suitability of a soil for certain crops.

Percent sodium is sometimes used. It is the ratio of sodium to the total cations in milliequivalents. A high sodium percentage may indicate a poor quality water, but in recent years more refinement in interpretation of water quality has made this less useful.

SOIL PROPERTIES AND WATER QUALITY

Over a long period of time the quality of irrigation water and the irrigation practice reach an equilibrium with the soil.

The Cations

To a great extent, the cations determine the physical as well as the chemical properties of soil. The cations of most concern are calcium, magnesium, sodium and potassium.

Calcium (Ca) is found in essentially all natural waters. A soil well supplied with exchangeable calcium is friable and easily tilled and usually permits water to penetrate readily. For these reasons, calcium in the form of gypsum is often applied to "tight" soils to

---

* Formerly referred to as millimhos per centimeter (mmhos/cm).

improve the physical properties. Calcium replaces the sodium on the
soil particles so that the sodium may be leached below the root zone.
Generally, an irrigation water containing predominantly calcium
is most desirable.

Magnesium (Mg) is also usually found in measurable amounts.
Magnesium behaves much like calcium in the soil. Often labora-
tories will not separate calcium and magnesium but will report simply
Ca + Mg in me/l. For most purposes this is adequate.

Sodium (Na) salts are all very soluble and as a result generally
are found in all natural waters.

A soil with a large amount of sodium associated with the clay
fraction has poor physical properties for plant growth. When it is
wet, it runs together, becomes sticky and is nearly impervious to
water. When it dries, hard clods form, making it difficult to till.
Continued use of water with a high proportion of sodium may bring
about these severe changes in an otherwise good soil. Eventually
these soils may become alkali. The so-called "slick spots" are usually
spots high in exchangeable sodium.

Potassium (K) is usually found in only small amounts in natural
waters. It behaves much like sodium in the soil. It is not often
reported separately in water analyses but is included with the
sodium.

The Anions

The anions indirectly affect the physical properties of soil by
altering the ratio of calcium and sodium attached to the clays. The
important anions are bicarbonate, carbonate, chloride and sulfate.

Bicarbonate (HCO₃) is common in natural waters. It is not usually
found in nature except in solution. Sodium and potassium bicarbonates can exist as solid salts, for example, baking soda (sodium
bicarbonate). Calcium and magnesium bicarbonates exist only in
solution. As the moisture in the soil is reduced, either by removal
by plants or by evaporation, calcium bicarbonate decomposes, carbon
dioxide (CO_2) goes off into the air and water (H_2O) is formed,
leaving insoluble lime ($CaCO_3$) behind.

$$Ca(HCO_3)_2 \xrightarrow{\text{upon drying}} CaCO_3 + CO_2 + H_2O$$

A similar reaction takes place with magnesium bicarbonate.
Large amounts of bicarbonate ions in irrigation water will, as the

soil approaches dryness, precipitate calcium, thereby removing it from the clay. This leaves sodium in its place. In this way, a calcium-dominant soil can become a sodium-dominant (sodic) soil by the use of a high bicarbonate irrigation water.

Carbonate (CO₃) is found in some waters. Since calcium and magnesium carbonates are relatively insoluble, high carbonate waters mean that the cations associated with them are likely to be sodium with possibly a small amount of potassium. Upon drying in the soil, the carbonate ion will remove calcium and magnesium from the clay similar to bicarbonate, and an alkali (sodic) soil will develop.

Chloride (Cl) is found in all natural waters. In high concentrations it is toxic to some plants. All common chlorides are soluble and contribute to the total salt content (salinity) of soils. The chloride content must be determined to properly evaluate irrigation waters.

Sulfate (SO₄) is abundant in nature. Sodium, magnesium and potassium sulfates are readily soluble. Calcium sulfate (gypsum) has a limited solubility. Sulfate has no characteristic action on the soil except to contribute to the total salt content. The presence of soluble calcium will limit its solubility.

Nitrate (NO₃) is not commonly found in large amounts in natural waters. Small amounts can affect its use as irrigation water by supplying plants with needed nitrogen, or in some cases with more than the desired amount of this plant nutrient. Large amounts of nitrates in water may indicate contamination from natural deposits and from many other sources such as animal wastes, sewage and decomposed soil organic matter. Nitrates, however, have no significant effect on the physical properties of soil.

Boron (B) occurs in water in one or another anion form. The usual range in natural waters is from 0.01 ppm to 10 ppm boron. Concentrations greater than this are known but are most often from hot springs or brines.

Boron has no measurable effect on the physical properties of soil in the amounts that can be tolerated by plants. Boron is not as readily removed from the soil as chloride or nitrate, but most of it can be removed by successive leachings.

A small amount of boron is essential for plant growth, but a concentration slightly above the optimum is toxic to plants. Some plants are more sensitive to an excess than others. Plants grown on some sandy soils which have been irrigated for several years by water exceedingly low in boron (less than 0.02 ppm) may develop boron deficiency.

EVALUATING IRRIGATION WATER

An ideal classification scheme should predict the effect of irrigation water on soils and plant growth. Many methods have been devised—all have been useful if not applied too broadly, but none have been completely satisfactory. It is doubtful if any system can be completely satisfactory without knowing how much water is applied, how much is leached below the root zone, the sensitivity of the crops to the various dissolved minerals, the frequency of irrigation, soil properties and even climatic conditions.

A *saline soil* is a soil containing soluble salts in such quantities that they interfere with the growth of crops. A saline soil is a non-alkali soil.

An *alkali soil* is a soil which contains enough sodium attached to the clay particles to interfere with the growth of crops. If an alkali soil is relatively free of soluble salts, it is called a non-saline alkali soil. If, in addition to being alkali, it has sufficient soluble salts to restrict plant growth, it is called a saline-alkali soil.

Salinity hazard—One of the hazards of irrigated agriculture is the possible accumulation of soluble salts in the root zone. Some

Fig. 2-2. Salts accumulate in the soil with inadequate leaching and drainage.

plants can tolerate more salts than others, but all plants have a maximum tolerance. With reasonably good irrigation practices the salt content of the saturation extract of soil is 1.5 to 3 times the salt content of the irrigation water. Where ample water is used to remove excess salt from the root zone, the salt level in the saturation extract is about 1.5 times that of the irrigation water. Where water is used more sparingly, there may be 3 times as much salt.

An acre-foot of water (the amount of water covering one acre, one foot deep) weighs approximately 2,720,000 pounds; therefore, 1 ppm of a salt in an acre-foot of water weighs 2.72 pounds. This means that one acre-foot of water containing only 735 ppm (EC = 1.15 ms/cm) carries *one ton of salt!* Many growers apply 4 feet of irrigation water per year to produce a crop—or they apply 4 tons of salt on every acre every year! This points out the need for adequate leaching below the root zone.

With ordinary irrigation methods there is some leaching, hence, the accumulation of salts in the soil water is reduced but not eliminated. Before a critical assessment of the salinity hazard of any irrigation water is made, it is necessary to know how much salt the crop can tolerate and how much leaching is needed to maintain the desired salt level in the soil water.

Tables 2-4, 2-5, 2-6 and 2-7 show the crop tolerance and leaching requirements estimated for several crops. Leaching need not be done at each irrigation but should be done at least once a year.

Growers rotating crops must provide enough leaching so that damage to the most salt-sensitive crop in the rotation will be at a minimum.

With reasonable irrigation practices, there should be no salinity problems with irrigation water with an EC of less than 0.75 ms/cm. Increasing problems can be expected between EC 0.75 and 3.0 ms/cm. An EC greater than 3.0 will cause severe problems except for areas restricted to only a few salt-tolerant crops.

It has generally been assumed that the effects of a saline water could be offset by increasing the amount of leaching so that the *average* salt content of the root zone would not be increased. The USDA Salinity Laboratory has demonstrated that yields of alfalfa (and probably other crops) are governed not by average soil salinity but primarily by the salinity of the irrigation water. Yields were reduced as the salinity of the water increased, no matter how much leaching was done. With this in mind, the practice of blending saline drain waters with low-salt irrigation waters needs to be reassessed.

Sodium or permeability hazards—In most cases permeability of

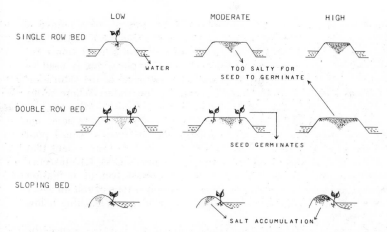

Fig. 2-3. Modification of seedbeds permits germination of seeds for good stand establishment.

water becomes a hazard before sodium as such affects plant growth. In a few plants, notably avocados, this may not be strictly true.

As the proportion of sodium attached to the clay in the soil increases, the soil tends to disperse or "run together" bringing about reduced rates of water penetration. The sodium adsorption ratio (SAR) indicates the relative activity of sodium ions as they react with clay. From the SAR the proportion of sodium on the clay fraction of the soil can be estimated when an irrigation water has been used for a long period of time with reasonable irrigation practice.

Most laboratories will report the SAR of irrigation waters. If not, it can be simply determined by using the following equation:

$$SAR = \frac{Na}{\sqrt{\dfrac{Ca + Mg}{2}}}$$

The sodium (Na), calcium (Ca) and magnesium (Mg) are expressed in me/l.

A refinement of the SAR called the "Adjusted SAR" (SAR adj) is now commonly used. SAR adj includes the added effects of precipitation and solution of calcium in soils as related to $CO_3 + HCO_3$ concentrations.

$$SAR \text{ adj} = \frac{Na}{\sqrt{\dfrac{Ca + Mg}{2}}} [1 + (8.4 - pHc)]$$

SAR adj = SAR $[1 + (8.4 - pHc)]$

pHc is a calculated value based on total cations, Ca + Mg and CO_3 + HCO_3 in the water. (For tables to calculate pHc, see the Wilcox reference at the end of the chapter.)

If the SAR adj is less than 6, there should be no problems with either sodium or permeability. In the range of 6 to 9 there are increasing problems, above 9, severe problems can be expected.

The SAR adj can be reduced by:

1. Increasing the calcium content of the water by adding gypsum or some other soluble calcium salt.
2. Reducing the HCO_3 in the water by adding sulfuric acid, sulfur dioxide (SO_2) or some other acidifying amendment.

SAR adj is a good index of the sodium and permeability hazard if the water passes through the soil and reaches equilibrium with it. Unfortunately, in the field this is not always the case. Water with relatively high bicarbonate and low calcium and magnesium content would have a high SAR adj and hence an exceedingly slow permeability. In many cases, plant growth will not permit water to remain on the crop long enough to allow for deep leaching. As transpiration continues, the salts are concentrated and this may lead to the development of a sodic soil.

Irrigation water with a very low salt content may also present a water penetration problem. In this case the addition of some salt, preferably a calcium salt such as gypsum, would be helpful. There is good evidence that all irrigation waters should contain a minimum of 20 ppm of calcium (1.0 me/l) to prevent dispersion of the soil.

TOXIC CONSTITUENTS

There are several elements found in natural waters which can be toxic to plants. Boron, chlorides and sodium are the ones commonly found. Waters high in bicarbonates have been shown to induce iron deficiencies in some plants, but this is minor when compared to their role in creating permeability problems.

Boron Hazard

A small amount of boron is necessary for plant growth. To sustain an adequate supply of this plant nutrient, 0.02 ppm B or more in the irrigation water may be required. Most waters contain an adequate supply of boron, but water from a few rivers may be deficient. Some well waters and a few surface streams contain an excess of boron, thus creating a health hazard by their use. Table 2-2 provides a satisfactory guide to the boron hazard in the irrigation water.

Plants grown in soils high in lime may tolerate more boron than those grown in non-calcareous soils.

Table 2-2.   Relative Tolerance of Plants to Boron*

Sensitive†	Semi-tolerant†	Tolerant†
(0.5 ppm)	(1 ppm)	(2 ppm)
Lemon	Lima bean	Carrot
Grapefruit	Sweet potato	Lettuce
Avocado	Bell pepper	Cabbage
Orange	Tomato	Turnip
Thornless blackberry	Pumpkin	Onion
Apricot	Zinnia	Broad bean
Peach	Oat	Gladiolus
Cherry	Milo	Alfalfa
Persimmon	Corn	Garden beet
Kadota fig	Wheat	Mangel
Grape (Sultanina and Malaga)	Barley	Sugar beet
Apple	Olive	Palm (Phoenix
Pear	Ragged robin rose	canariensis)
Plum	Field pea	Date palm
American elm	Radish	(Phoenix dactylifera)
Navy bean	Sweet pea	Asparagus
Jerusalem artichoke	Pima cotton	Athel (Tamarix aphylla)
Persian (English) walnut	Acala cotton	
Black walnut	Potato	
Pecan	Sunflower (native)	
(1 ppm)	(2 ppm)	(4 ppm)

* Adapted from USDA Technical Bulletin 448. Number values apply to irrigation waters. For soils, the number values may be somewhat higher.
† The plants first named are considered more sensitive and the last named more tolerant: 0.5 ppm, satisfactory for all crops; 0.5-1.0 ppm, satisfactory for most crops, with some leaf injury on sensitive crops; 1.0-2.0 ppm, satisfactory for some crops, with sensitive crops reduced in yield and vigor; 2.0-4.0 ppm, satisfactory for only tolerant crops; >4.0 ppm, unsatisfactory for all crops.

Chloride Hazard

Chlorides are found in all natural waters. The common chlorides are soluble and are not fixed in the soil so that they can move through the soil and into the drainage water. Chlorides in relatively small amounts are necessary for plant growth. In high concentrations, however, chlorides will inhibit plant growth, and they are specifically toxic to some plants.

Most annual crops and short-lived perennials are moderately to highly tolerant to chlorides, and growers can rely on the salinity hazard index for evaluating water use problems. Trees, vines and woody ornamentals are sensitive to chlorides. For these plants, Table 2-3 may serve as a guide.

Table 2-3.   Chloride Toxicity Index

Chlorides (me/l)	Chlorides (ppm)	Notes
Less than 2	Less than 70	Generally safe for all plants.
2 - 4	70 - 140	Sensitive plants usually show slight to moderate injury.
4 - 10	140 - 350	Moderately tolerant plants usually show slight to substantial injury.
Greater than 10	Greater than 350	Can cause severe problems.

Sodium Toxicity

Trees, vines and woody ornamentals may be sensitive to excessive sodium absorbed through the plant roots. As in the case of chlorides, annual crops are usually not affected except as a contribution to total salt content. For waters with SAR adj below 3 there is no problem; from SAR adj values of 3 to 9, problems increase, and above 9 they are severe. For waters this high, severe permeability problems would also exist.

WATER FOR SPRINKLER IRRIGATION

Most water suitable for surface irrigation may be safely used for overhead sprinkler irrigation. There are, however, some exceptions.

Leaf burn caused by sodium and chloride absorption may occur when the rate of evaporation is high. Conditions such as low humidity, high temperature and winds can increase the concentration of these ions in the water on the leaves between rotations of the sprinklers. Sometimes this can be corrected by increasing the rate of rotation. If this is impractical, it may be necessary to irrigate only at night during periods of hot, dry weather. Usually there is no problem when the irrigation water contains 3 me/l or less of either sodium or chloride.

Bicarbonate ions in water can also be a problem with overhead sprinkler irrigation. A white deposit of calcium carbonate forms on the leaves and fruit. This can render some fruits and ornamentals unmarketable because they are unattractive. This coat of "whitewash" is not known to have any other adverse effect on plant growth. Levels below 1.5 me/l of HCO_3 should cause no problem.

DRIP IRRIGATION

Drip irrigation is the frequent slow application of water through various types of emitters. Drip systems vary according to the crop, for example, a mature tree may require two to six emitters, each emitter applying water at the rate of $\frac{1}{2}$ to 2 gallons per hour.

These low rates of application require small outlets and therefore need water which has been filtered free of solid particles. Dissolved salts in the water may crystallize around the outlets and reduce their size or plug them completely.

The total amount of water used in drip irrigation is usually less than for conventional irrigation systems because:

1. Tailwater runoff is completely eliminated.
2. Evaporation from the soil surface is reduced because less of the soil surface is wetted.
3. The total volume of soil wetted usually is less.
4. Deep percolation of water may be reduced.

Drip irrigation is mose adapted to permanent crops such as orchards and vineyards, especially on irregular slopes. Since it may save water, it is also adapted to areas of costly or scarce water.

Drip irrigation systems are used on some annual crops where costs can be justified.

CROP SALINITY TOLERANCE AND LEACHING REQUIREMENT

The following crop tolerance tables show:

1. The yield reduction to be expected due to salinity of irrigation water (EC_w).
2. The salt content of the soil saturation extract (EC_e) when common surface irrigation methods are used.
3. The leaching requirement (LR), which is the fraction of the irrigation water that must be leached through the active root zone to control salinity at a specified level.
4. The maximum concentration of salts that the plant can tolerate (EC_e Max).

These tables are from data developed by the USDA Salinity Laboratory, Riverside, California.

An example of how to use these tables. Assume the crop is *corn*.

From the tables:

Maximum EC_{dw} = 10 ms/cm

Maximum EC_w:

for   0% yield loss = 1.1 ms/cm with LR 6%
      10%   "   "   = 1.7 ms/cm with LR 8%
      25%   "   "   = 2.5 ms/cm with LR 12%
      50%   "   "   = 3.9 ms/cm with LR 20%

Suppose the water has an EC_w equaling 2.7. Using the common surface irrigation methods, one can expect the EC_e of the most active root zone to be 1.5 times the EC_w (2.7) or 4.0 ms/cm. The tables show that a decrease in yield of slightly less than 25 percent can be expected.

To maintain the EC_e at the base of the roots below the maximum safe level, some water must pass below the root zone. This minimum leaching requirement may be calculated by the following equation:

$$LR = \frac{EC_w}{2^* \times EC_e \text{ Max}} \times 100 = \frac{2.7 \times 100}{2 \times 10} = 13.5\%$$

SALT MOVEMENT IN SOIL

Soluble salts in the soil move in the direction of water movement. In areas of overall flooding or sprinkling, salt movement is directly downward. Movement and concentration of salts on beds, border checks or berms in orchards or vineyards commonly occur.

For most crops, seedlings are more sensitive to soluble salts in the soil than are established plants. Planting on the shoulder of the bed

---

* EC of drainage water is twice the EC_e (extract of saturated soil).

Table 2-4. Field Crops—Reduction in Yield*

Crop	ECe†	ECw‡	LR	ECe	ECw	LR	ECe	ECw	LR	ECe	ECw	LR	ECe§
	(0%)			(10%)			(25%)			(50%)			(Maximum)
Barley||	8.0	5.3	9%	10.0	6.7	12%	13.0	8.7	16%	18.0	12.0	21%	28.0
Cotton	7.7	5.1	9%	9.6	6.4	12%	13.0	8.4	16%	17.0	12.0	22%	27.0
Sugar beets#	7.0	4.7	10%	8.7	5.8	12%	11.0	7.5	16%	15.0	10.0	21%	24.0
Wheat||,**	6.0	4.0	10%	7.4	4.9	12%	9.5	6.4	16%	13.0	8.7	22%	20.0
Safflower	5.3	3.5	12%	6.2	4.1	14%	7.6	5.0	17%	9.9	6.6	23%	14.5
Soybeans	5.0	3.3	16%	5.5	3.7	18%	6.2	4.2	21%	7.5	5.0	25%	10.0
Sorghum	4.0	2.7	8%	5.1	3.4	9%	7.2	4.8	13%	11.0	7.2	20%	18.0
Rice (paddy)	3.0	2.0	9%	3.8	2.6	11%	5.1	3.4	15%	7.2	4.8	21%	11.5
Sesbania	2.3	1.5	5%	3.7	2.5	8%	5.9	3.9	12%	9.4	6.3	19%	16.5
Corn	1.7	1.1	6%	2.5	1.7	8%	3.8	2.5	12%	5.9	3.9	20%	10.0
Flax	1.7	1.1	6%	2.5	1.7	8%	3.8	2.5	12%	5.9	3.9	20%	10.0
Cowpeas	1.3	0.9	5%	2.0	1.3	8%	3.1	2.1	12%	4.9	3.2	19%	8.5
Beans (field)	1.0	0.7	5%	1.5	1.0	8%	2.3	1.5	12%	3.6	2.4	18%	6.5

* Adapted from "Quality of Water for Irrigation." R. S. Ayers, Jour. of the Irrig. and Drain. Div., ASCE. Vol. 103, No. IR2, June 1977, p. 140.

† ECe means electrical conductivity of the saturation extract of the soil reported in ms/cm at 25° C.
‡ ECw means electrical conductivity of the irrigation water in ms/cm at 25° C.
§ Maximum ECe is the electrical conductivity of the soil saturation extract at which crop growth ceases.
\|\| Barley and wheat are less tolerant during germination and seedling stage, ECe should not exceed 4 or 5 ms/cm.
Sensitive during germination. ECe should not exceed 3 ms/cm for garden beets and sugar beets.
** Tolerance data may not apply to new semi-dwarf varieties of wheat.

Table 2-5. Vegetable Crops—Reduction in Yield*

Crop	(0%) EC_e	EC_w	LR	(10%) EC_e	EC_w	LR	(25%) EC_e	EC_w	LR	(50%) EC_e	EC_w	LR	EC_e (Maximum)
Beets†	4.0	2.7	9%	5.1	3.4	11%	6.8	4.5	15%	9.6	6.4	21%	15.0
Broccoli	2.8	1.9	7%	3.9	2.6	10%	5.5	3.7	14%	8.2	5.5	20%	13.5
Tomatoes	2.5	1.7	7%	3.5	2.3	9%	5.0	3.4	14%	7.6	5.0	20%	12.5
Cantaloupes	2.2	1.5	5%	3.6	2.4	8%	5.7	3.8	12%	9.1	6.1	19%	16.0
Cucumbers	2.5	1.7	8%	3.3	2.2	11%	4.4	2.9	14%	6.3	4.2	21%	10.0
Spinach	2.0	1.3	4%	3.3	2.2	7%	5.3	3.5	12%	8.6	5.7	19%	15.0
Cabbage	1.8	1.2	5%	2.8	1.9	8%	4.4	2.9	12%	7.0	4.6	19%	12.0
Potatoes	1.7	1.1	6%	2.5	1.7	9%	3.8	2.5	13%	5.9	3.9	20%	10.0
Sweet corn	1.7	1.1	6%	2.5	1.7	9%	3.8	2.5	13%	5.9	3.9	20%	10.0
Sweet potatoes	1.5	1.0	5%	2.4	1.6	8%	3.8	2.5	12%	6.0	4.0	19%	10.5
Peppers	1.5	1.0	6%	2.2	1.5	9%	3.3	2.2	13%	5.1	3.4	20%	8.5
Lettuce	1.3	0.9	5%	2.1	1.4	8%	3.2	2.1	12%	5.2	3.4	19%	9.0
Radishes	1.2	0.8	4%	2.0	1.3	7%	3.1	2.1	12%	5.0	3.4	19%	9.0
Onions	1.2	0.8	5%	1.8	1.2	8%	2.8	1.8	12%	4.3	2.9	19%	7.5
Carrots	1.0	0.7	4%	1.7	1.1	7%	2.8	1.9	12%	4.6	3.1	19%	8.0
Beans	1.0	0.7	5%	1.5	1.0	8%	2.3	1.5	12%	3.6	2.4	18%	6.5

* Adapted from "Quality of Water for Irrigation." R. S. Ayers, Jour. of the Irrig. and Drain. Div., ASCE. Vol. 103, No. IR2, June 1977, p. 141.
† Sensitive during germination. ECe should not exceed 3 ms/cm for garden beets and sugar beets.

Table 2-6. Fruit and Nut Crops—Reduction in Yield*

Crop	(0%) EC_e	EC_w	LR	(10%) EC_e	EC_w	LR	(25%) EC_e	EC_w	LR	(50%) EC_e	EC_w	LR	EC_e (Maximum)
Date palms	4.0	2.7	4%	6.8	4.5	7%	10.9	7.3	11%	17.9	12.0	19%	32.0
Figs, Olives, Pomegranates	2.7	1.8	6%	3.8	2.6	9%	5.5	3.7	13%	8.4	5.6	20%	14.0
Grapefruit	1.8	1.2	8%	2.4	1.6	10%	3.4	2.2	14%	4.9	3.3	21%	8.0
Oranges	1.7	1.1	7%	2.3	1.6	10%	3.2	2.2	14%	4.8	3.2	20%	8.0
Lemons	1.7	1.1	7%	2.3	1.6	10%	3.3	2.2	14%	4.8	3.2	20%	8.0
Apples, Pears	1.7	1.0	6%	2.3	1.6	10%	3.3	2.2	14%	4.8	3.2	20%	8.0
Walnuts	1.7	1.1	7%	2.3	1.6	10%	3.3	2.2	14%	4.8	3.2	20%	8.0
Peaches	1.7	1.1	8%	2.2	1.4	11%	2.9	1.9	15%	4.1	2.7	21%	6.5
Apricots	1.6	1.1	9%	2.0	1.3	11%	2.6	1.8	15%	3.7	2.5	21%	6.0
Grapes	1.5	1.0	4%	2.5	1.7	7%	4.1	2.7	11%	6.7	4.5	19%	12.0
Almonds	1.5	1.0	7%	2.0	1.4	10%	2.8	1.9	14%	4.1	2.7	19%	7.0
Plums	1.5	1.0	7%	2.1	1.4	10%	2.9	1.9	14%	4.3	2.8	20%	7.0
Blackberries	1.5	1.0	8%	2.0	1.3	11%	2.6	1.8	15%	3.8	2.5	21%	6.0
Boysenberries	1.5	1.0	8%	2.0	1.3	11%	2.6	1.8	15%	3.8	2.5	21%	6.0
Avocados	1.3	0.9	8%	1.8	1.2	10%	2.5	1.7	14%	3.7	2.4	20%	6.0
Raspberries	1.0	0.7	6%	1.4	1.0	9%	2.1	1.4	13%	3.2	2.1	19%	5.5
Strawberries	1.0	0.7	9%	1.3	0.9	11%	1.8	1.2	15%	2.5	1.7	21%	4.0

* Adapted from "Quality of Water for Irrigation." R. S. Ayers, Jour. of the Irrig. and Drain. Div., ASCE. Vol. 103, No. IR2, June 1977, p. 142.

Table 2-7. Forage Crops—Reduction in Yield*

Crop	EC_e	EC_w	LR	EC_e	EC_w	LR	EC_e	EC_w	LR	EC_e	EC_w	LR	EC_e
	(0%)			(10%)			(25%)			(50%)			(Maximum)
Tall wheat grass	7.5	5.0	8%	9.9	6.6	10%	13.3	9.0	14%	19.4	13.0	21%	31.5
Wheat grass (fairway)	7.5	5.0	11%	9.0	6.0	14%	11.0	7.4	17%	15.0	9.8	22%	22.0
Bermudagrass	6.9	4.6	10%	8.5	5.7	13%	10.8	7.2	16%	14.7	9.8	22%	22.5
Barley (hay)†	6.0	4.0	10%	7.4	4.9	12%	9.5	6.3	16%	13.0	8.7	22%	20.0
Perennial ryegrass	5.6	3.7	10%	6.9	4.6	12%	8.9	5.9	16%	12.2	8.1	21%	19.0
Birdsfoot trefoil, narrow leaf	5.0	3.3	11%	6.0	4.0	13%	7.5	5.0	17%	10.0	6.7	22%	15.0
Harding grass	4.6	3.1	9%	5.9	3.9	11%	7.9	5.3	15%	11.1	7.4	21%	18.0
Tall fescue	3.9	2.6	6%	5.8	3.9	8%	8.6	5.7	12%	13.3	8.9	19%	23.0
Crested wheat grass	3.5	2.3	4%	6.0	4.0	7%	9.8	6.5	11%	16.0	11.0	19%	28.5
Vetch	3.0	2.0	8%	3.9	2.6	11%	5.3	3.5	15%	7.6	5.0	21%	12.0
Sudan grass	2.8	1.9	4%	5.1	3.4	7%	8.6	5.7	11%	14.4	9.6	18%	26.0
Big trefoil	2.3	1.5	10%	2.8	1.9	13%	3.6	2.4	16%	4.9	3.3	22%	7.5
Alfalfa	2.0	1.3	4%	3.4	2.2	7%	5.4	3.6	12%	8.8	5.9	19%	15.5
Clover, berseem	1.5	1.0	3%	3.2	2.1	6%	5.9	3.9	10%	10.3	6.8	18%	19.0
Orchardgrass	1.5	1.0	3%	3.1	2.1	6%	5.5	3.7	11%	9.6	6.4	18%	17.5
Meadow foxtail	1.5	1.0	4%	2.5	1.7	7%	4.1	2.7	11%	6.7	4.5	19%	12.0
Clover, alsike, ladino, red, strawberry	1.5	1.0	5%	2.3	1.6	8%	3.6	2.4	12%	5.7	3.8	19%	10.0

* Adapted from "Quality of Water for Irrigation," R. S. Ayers. Jour. of the Irrig. and Drain. Div., ASCE. Vol. 103, No. IR2, June 1977, p. 143.
† Barley and wheat are less tolerant during germination and seedling stage. ECe should not exceed 4 or 5 ms/cm.

or planting two rows on a wide bed so that the salts will be pushed to the center of the bed and away from the seeds or seedlings may help to prevent damage (see Figure 2-3). If either the soil or the irrigation water is marginally saline, irrigating in alternate furrows may be necessary (see Figure 2-4). Sprnkling irrigated seedlings until they become well established may also be desirable.

Excessive salts may accumulate in the tops of beds during pre-

PATTERNS OF SALT ACCUMULATION

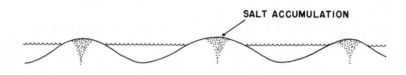

SALT ACCUMULATION

IRRIGATING EACH FURROW

IRRIGATING ALTERNATE FURROWS

BORDER CHECKS IN FIELD CROPS OR
BERMS IN VINEYARDS OR ORCHARDS

Fig. 2-4. Irrigation management can affect salt accumulation.

irrigation. This is particularly true where animal manures have been used and subsequent rainfall has not been great enough to remove excess salt before planting the seed. In this case, it may be necessary to sprinkle irrigate to remove the salt or to disperse the salts before seeding.

In areas of low rainfall, permanent berms in the rows of orchards and vineyards may accumulate excessive salts in a few years even where relatively low-salt water is used. In such cases, these berms should be removed to disperse the salt and then be rebuilt.

DRAINAGE

Most crops grow best where the water table is more than 6 feet below the soil surface. Fields where water stands within 6 feet of the surface should be drained. This can be accomplished by the installation of a tile or other drainage system. Tile will drain water

Fig. 2-5. Water makes the desert bloom. Photo by Valmont Industries, Inc.

only if placed below the water table or zone of saturation. Tile will not drain soils which are not saturated with water. If a field is poorly drained and there is an outlet for the drainage water, a carefully designed and installed drainage system may pay handsomely. These shallow waters are exceedingly saline in many areas and may not be suitable for irrigation.

If carefully managed, many crops can get a large share of their water needs from shallow ground water provided the ground water is not excessively saline.

SUPPLEMENTARY READING

1. *Diagnosis and Improvement of Saline and Alkali Soils.* USDA Agricultural Handbook No. 60. 1954.
2. "Crop Salt Tolerance—Current Assessment." E. V. Mass and G. J. Hoffman. Jour. of the Irrig. and Drain. Div., ASCE. 103: 115-134. June 1977.
3. "Irrigation Management for Salt Control." J. Van Schilfgaarde, L. Bernstein, J. D. Rhoades and S. L. Rawlins. Jour. of the Irrig. and Drain. Div., ASCE. 98: 322-338. 1972.
4. "Quality of Water for Irrigation." J. D. Rhoades. *Soil Sci.* 113: 277-284. April 1972.
5. "Quality of Water for Irrigation." R. S. Ayers. Jour. of the Irrig. and Drain. Div., ASCE. 103: 135-154. June 1977.
6. *Salt Tolerance of Fruit Crops.* USDA Agricultural Information Bulletin 262. 1965.
7. *Salt Tolerance of Plants.* USDA Agricultural Information Bulletin 263. 1964.
8. *Salt Tolerance of Vegetable Crops.* USDA Agricultural Information Bulletin 205. 1959.
9. *Tables for Calculating pHc Values of Waters.* L. V. Wilcox. USDA Salinity Lab. Mineo. December 1966.
10. "Water Quality for Agriculture." R. S. Ayers and D. W. Westcott. FAO. Rome. 1976.

Chapter 3

PRINCIPLES OF PLANT GROWTH

Webster defines growth as "a growing; increase; esp., progressive development of an organism, or the like." This is well illustrated by the process of planting a seed and initiating the process that results in the production of a whole plant. Plant growth is the process that provides man with his food supply and much of his shelter. Over a century ago the noted scientist Justus von Liebig said about plant growth: it is "the primary source whence man and animals derive the means of their growth and support."

Growth is fundamentally one of the attributes of living protoplasm. It is more than a mere increase in size and weight, although this may be one expression of it. Growth represents a progressive and irreversible change in form involving the formation of new cells and their enlargement and maturation into the tissues and organs of the plant.

All plants must have, in varying degrees, the same basic supply for growth of light, heat, energy, water, oxygen, carbon and mineral elements. Usually the soil supplies the needed moisture and mineral elements while the air supplies the oxygen and carbon dioxide. Growth is stopped, started or at least modified as temperature conditions change, both in the root zone and around the aerial portion of the plants. This chapter will discuss briefly the requirements for satisfactory growth and introduce some of the concepts involved in the mineral nutrition of plants.

THE PLANT CELL

All living plant parts are made up of cells. The single plant cell is the basic structural and functional unit of the plant. It is a tiny chemical factory that absorbs and secretes materials; transforms light energy into chemical energy (photosynthesizing cells); respires and releases energy for various activities; digests or transforms foods; synthesizes complex chemicals from air, water and simple sugars; contains and even synthesizes the remarkable substance called protoplasm. These are but a few of the activities carried on within the cell which are basic to the life process itself.

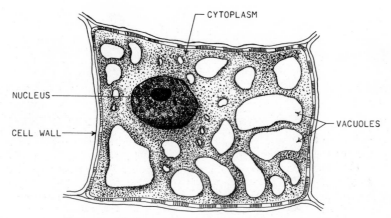

Fig. 3-1. All living plant parts are made up of cells.

PLANT TISSUES

Cells that are grouped into the same functional unit are referred to as tissues. Tissues may then be classified by function into one of four groups: meristematic, fundamental, protective and vascular. Meristematic tissues are the embryonic and undifferentiated cells, capable of growing by division and occurring at growing points. Fundamental tissues are made up of masses of cells which have little specialization in structure or function. They act primarily as storage units. The epidermal or "skin" surface of a plant usually contains protective tissues. Vascular tissues such as the xylem and the phloem function in the conductive processes of the plant. They are highly specialized tissues that also add mechanical support because of their structure and location.

PLANT ORGANS

The organization of a group of tissues forms an organ. Organs are generally separated into roots, leaves and stems, although throughout the plant kingdom there are plants that may have one or more of these organs missing and other specialized organs present. A careful study of botany will reveal many interesting things about plant life and contribute to one's understanding of its complexities.

ROOTS

The root is that part of the plant which ordinarily grows downward into the soil, anchoring the plant and absorbing water and mineral nutrients. It may also serve as a food storage organ or reproductive organ or it may perform some other function. It takes many physical forms and varies in extension from a few feet to many miles on a single plant. In a careful study of a single rye plant that was allowed to grow for four months in a box 12 inches square and 22 inches deep, it was found that the accumulated length of all the roots was 387 miles. The total area of the root surface of that one plant was 6,875 square feet.

The root differs from the shoot portion of the plant primarily in structure. Unlike stems, roots do not bear leaves or regular buds and are not divided into nodes and internodes. Usually, the root differs from the shoot in function and location, but this is not always so since some plants have roots that develop buds which give rise to leafy shoots, and other plants have aerial stems which absorb water and nutrients. Tubers and stolons are stems often found underground, while brace roots of corn and air roots of orchids are found above ground.

Root systems are often roughly grouped into two general types, fibrous and taproot. When numerous long, slender roots of about equal size are developed, they are known as fibrous roots. Examples of this root classification are corn, small grains and the grasses. If the primary root remains the largest root of the plant and continues its downward development with other roots developing from it, it is classified as a taproot. Examples of this are cotton, alfalfa, sugar beet, dandelion and ragweed. Root systems not well defined as fibrous or taproot are quite common also. Whatever the classification, it is important to know the nature of the root system of a plant if one is to know how to properly manage the growing of that plant. How extensive is the root system? How deep does it grow? How rapidly does it develop? No plant can achieve its optimum development with a poor root system.

Numerous environmental factors influence the direction and extent of root growth. These include light, gravity, temperature, salt concentration, soil texture and physical condition, oxygen supply, moisture and mineral nutrient supply. Each root becomes subject to a combination of all of these factors, and its growth is the resultant of the combined action of all of them.

Roots develop best in "loose" soil and in fertile soil since nitrogen and phosphorus seem to stimulate root development. Thus, one of the purposes of tillage and fertilization is to provide a proper environment for the establishment and development of a good root system.

Oxygen must be available to all living cells. The amount necessary for growth varies with species. Unless the plant has specially developed systems for transporting oxygen to the roots such as some of the swamp plants and rice, flooding of soils for appreciable lengths of time will cause death because of lack of oxygen.

Downward growth is the common response of roots to the pull of gravity. This is caused by unequal distribution of hormones inducing their effect on growth. Downward bending by the roots is known as positive geotropism, and upward bending by the shoots is negative geotropism.

The effect of light on hormonal concentration is to cause plants to bend towards the light. This again is caused by unequal hormonal distribution caused by the action of light. Bending towards light is termed positive phototropism; bending away from light is known as negative phototropism. Some roots exhibit negative phototropism.

Roots will generally bend in the direction of most favorable temperatures, thus exhibiting a positive thermotropism, and they will grow in the direction of favorable moisture supply, positive hydrotropism. These are responses to the immediate environment and are not some sensory perceptive seekings by the plant. Roots do not *seek* favorable temperatures or water, but grow in areas of favorable temperatures or moisture to which they are intimately exposed.

A close look at a root will show that it is made up of many different functioning parts. The primary absorbing region for water and mineral nutrients is the younger portion near the root tip. An even closer look in this region will reveal tiny root hairs radiating outward from all sides of the root. These root hairs are not much over 0.01 millimeter in diameter and a few millimeters in length. Each hair is an outward prolongation of a portion of an epidermal cell. The outer, delicate walls of the root hair consist partly of pectic materials which are gelatinous and enable the root hair to cling to soil particles and to absorb water and salts in solution. It is common to find 200 to 300 root hairs per square millimeter of epidermis in the root-hair zone.

Proceeding from the growing tip back to the older root zone, the roots take on more the appearance and, to a degree, the functions of underground stems. Their primary role becomes one of conducting water and mineral nutrients from the absorbing root tip zone and transporting synthesized compounds from the leaves to the root tips.

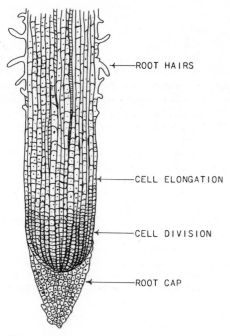

ROOT HAIRS

CELL ELONGATION

CELL DIVISION

ROOT CAP

Fig. 3-2. A microscopic view of a root tip.

As a rule most of the water and minerals the plant obtains are taken in by the younger roots in the newly developed root area where the root hairs are most numerous. The older tissues back of this region become progressively impermeable although it has been shown that there may be considerable water absorption through these less active regions of the root, particularly in tree crops.

The process of absorption is one of the main functions of the root. Without a constant supply of water, the plant cannot carry on the basic physiological activities such as photosynthesis, respiration and growth. Without a supply of mineral nutrients brought in by absorptive processes the plant would cease to live.

The study of plant physiology has brought out much concerning absorption of mineral nutrients. It has been shown to be a process requiring an expenditure of energy. For example, ion concentrations may be 1,000 times greater within the cell than in the soil solution immediately outside the cell.

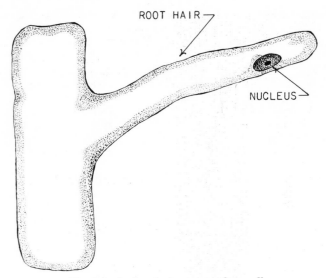

Fig. 3-3. A close look at a root-hair cell.

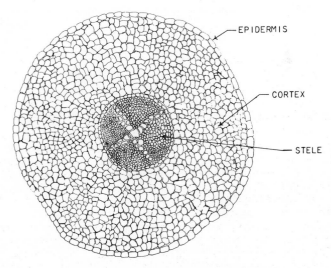

Fig. 3-4. A cross section of a typical root.

A brief, greatly simplified description of the absorption and move-ment of a mineral ion such as potassium will serve to illustrate the absorptive process. A typical young plant root cell consists of cyto-plasm encircled on the outside by a cell wall. Within the cytoplasm are vacuoles or kinds of voids holding organic solutes and inorganic ions in solution. The membrane adjoining the cell wall is termed the *plasmalemma,* while another membrane termed the *tonoplast* forms the boundary between the vacuole and the cytoplasm. It is from the outer cell wall, through the plasmalemma, across the cytoplasm, the tonoplast and into the vacuole that the potassium ion must pass. Then for it to move into the conductive xylem tissue, it must traverse other cells and pass through additional membranes.

The potassium ion diffuses from the soil to the root surface. Here it freely moves through the cell wall as it is carried in the soil solution. At this point it contacts the plasmalemma. It is this membrane which is highly impermeable to ions that control the further movement of the potassium ion into the root. Certain locations or sites are found in this membrane which are specific for specific ions. At this point it is theorized that a carrier system attaches itself to the potassium ion, transports it across the plasmalemma and deposits it on the other side. The ion is thus held inside by the membrane, and the carrier is regenerated to pick up another ion.

Once inside the cell the potassium ion can then move by diffusion, by mass flow in the transpiration stream or by processes regulated by metabolism. Entry into the upward moving stream focused primarily in the xylem tissues allows for rather rapid movement of the potas-sium ions throughout the whole plant.

Water movement is primarily a physical process. As water evapor-ates from leaves it creates a difference in tension between the leaves and roots. This tension "pulls" water up the plant. Since the initial entry of water into the root is mainly through the active region of the root, into and through the cell walls, it is quite like a long tube. The major impediment to the movement is through the endodermis where it must pass through cytoplasmic space. This is the area that is influenced by metabolism and which becomes subject to factors which affect metabolism such as temperature, oxygen supply, meta-bolic poisons, carbohydrate or food supply, etc.

Once water has entered the major conductive tissues it travels with relatively minimal resistance to flow. These tissues are the xylem and phloem. They are usually adjacent to one another. They con-stitute the vascular tissues of the roots and the stems.

Numerous books have been written concerning such subjects as mineral nutrition of plants, plant and soil water relationships and water relations of plants. Also, hundreds of scientific articles have been written on these subjects. A few helpful references are included at the end of this chapter.

SHOOTS

The leaf and the stem make up the shoot. Each of these structures has separate but often overlapping functions. In general the leaf is the center of photosynthetic activity—the food manufacturing process. Its structure facilitates this function as it is physically positioned to get maximum exposure to the sun. Along its surface are minute openings called stomata where exchange of carbon dioxide and oxygen occurs. Underneath the surface or epidermal layer are specialized cells called the spongy layer, which are ideally suited to perform the photosynthetic process. Here the gases from the air, water in the cells and light from the sun all join in the presence of chloroplasts, the chlorophyll-bearing bodies, to perform the photosynthetic process.

The stems of some plants also carry on photosynthesis although these structures usually are the conductive tissues that carry water and mineral nutrients to the leaves and photosynthesized food to the

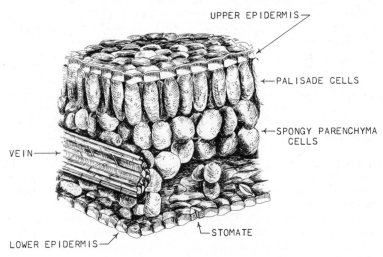

UPPER EPIDERMIS

PALISADE CELLS

SPONGY PARENCHYMA CELLS

VEIN

STOMATE

LOWER EPIDERMIS

Fig. 3-5. A cross section view of a leaf.

roots or other organs. They also serve to display leaves and may also become the food storage and reproductive units of the plant.

PHOTOSYNTHESIS

The manufacture of food substances from carbon dioxide, water and mineral elements in the presence of light and chlorophyll in living green plant tissues is photosynthesis. This is simply expressed in chemical equation form as:

$$6\,CO_2 + 12\,H_2O \xrightarrow[\text{chlorophyll}]{\text{light}} C_6H_{12}O_6 + 6\,O_2 + 6\,H_2O$$

carbon dioxide   water   sugar   oxygen   water

The process of respiration is just the reverse, or simply stated, the consumption of sugar in the presence of oxygen and water. In this case, light and chlorophyll are not needed.

The sugar produced by photosynthesis is transported to other plant parts. There it may be converted to starch, proteins, fats and other compounds or it may be respired. It is the photosynthetic process which provides the animal world with its food and also serves to make food available for future plant growth or reproduction.

TRANSPIRATION

The evaporation of water from leaves, stems and other aerial parts of plants is called transpiration. As much as 99 percent or more of the water absorbed by the roots may be transpired, yet the amount utilized is vital to the life processes of the plant. That portion evaporating has a cooling effect and serves to create a pressure pull within the conductive tissues which helps to explain the movement of solutes and other materials in the plant's vascular system.

External factors such as light, temperature, humidity, wind velocity and soil moisture conditions influence the rate of transpiration in plants. The minute openings in the leaves called stomata open in response to light, thus, transpiration is much greater in the daytime than at night. Response by the stomata to opening and closure is related to changes within the living cells which are affected by external and internal factors.

Modification in plant structure is a natural process which governs plant adaptability to climatic factors. Waxy cuticles, thickened layers, decreased leaf surface, fewer stomata and hairy leaf surfaces are natural systems to modify transpiration rate and allow for greater resistance to moisture stress.

A quantity factor has been developed to express the ratio of total water absorbed to the total amount of dry matter produced by the plant. This is termed the "water requirement" or "transpiration ratio."

The water requirement of an adequately fertilized crop is essentially the same as one inadequately supplied with nutrients. This indicates a reduction in water required per unit of production and a better water use efficiency.

FACTORS AFFECTING GROWTH

How fast a plant grows and the shape or form it assumes is determined by internal and external factors. Heredity is the predominant internal factor, and environment is the major external factor.

Heredity

The tendency for an offspring to display the characteristics of its parents is known as heredity. If we consider a wheat plant, the length and strength of the stem, the shape and texture of the leaf, the number of spikelets in the head, the shape, color and surface of the glumes, the weight of the kernel, the character of the grain, the yield of the seed, the resistance to cold, drought and disease, together with many other characteristics, have all been shown to be inheritable. The male and female sexual cells (gametes) which unite to produce the offspring contain these specific characteristics, and the offspring is a product of that union.

The genes are the remarkable minute pieces of protoplasm located in the chromosomes of the cells that carry the genetic characteristics of the organism. These genes provide the map or the blueprint for all developing cells. Whether the plant is tall or short, the leaf is round or pointed, the flower is red or white, in short, the very genetic makeup of the plant is established by the pattern set by the combination of genes.

Mendel is known as the father of plant genetics since he set down the theory of inheritance and demonstrated that theory. Using the garden pea he was able to single out a particular characteristic, such as flower color or smoothness of the seed, and demonstrate how this characteristic was inherited. He kept accurate records of the crosses he made, knew the ancestry of each individual and found the number of individuals in the offspring that followed the characteristics of the parents. Although the science of genetics is much more complex today, the painstaking and careful scientific methods established by Mendel are still followed in genetic experimentation.

Growth vs. Time

The growth of a cell, an organ or a whole plant does not proceed at a uniform rate. Growth starts out slowly, gradually increasing until a maximum rate is reached and then slows until it ceases altogether. A characteristically S-shaped curve is obtained if total growth is plotted against time. This is illustrated in Figure 3-6.

Fluctuations in temperature, moisture supply or other environmental conditions may cause irregularities in the curve, but if the whole period of growth is considered, the shape of the curve will remain the same. When nutrient uptake is plotted against time, the

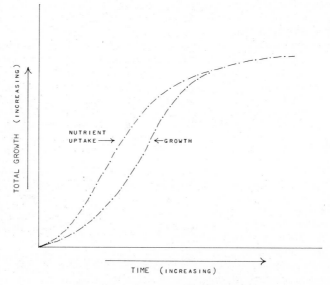

Fig. 3-6. A characteristic growth curve.

accumulation of nutrients closely follows the growth curve shape. Note that it precedes the growth since the nutrients must be present for growth to occur. A temporary shortage of nutrients will cause irregularities in the curve, and a severe shortage will essentially stop the growth.

Temperature

Temperature has a marked effect upon plant growth and is one of the most important factors determining the distribution of plants over the earth's surface. The temperatures within which plants are able to grow are often designated as minimum, optimum and maximum. The points below and beyond which growth ceases are designated minimum and maximum, while the temperature at which growth proceeds best is termed optimum. These points are not fixed during the life of the plant nor are they the same for all parts of the plant. In general, for temperate climates, the minimum usually falls somewhere above freezing, optimum at about 80° to 90°F and maximum around 110° to 120°F.

The direct effect of temperature must be evaluated in terms of its effect on such basic processes as photosynthesis, respiration, water and nutrient absorption, chemical processes within the plant, etc. Also, temperature has an effect upon the soil and chemical transformation within the soil. These are intimately related to root activity as illustrated in reduced uptake of phosphorus from a cool soil. Microbial activity is accelerated as temperature of the soil reaches optimum and results in the release of mineral nutrient elements from soil organic matter and plant residues.

Radiant Energy

The amount, quality and duration of sunlight play an important part in plant growth and development. Most plants are able to reach maximum growth at less than full sun intensity; however, this is often modified by density of plant canopy and shading. Some plants are better able to use the maximum sunlight. Genetic characteristics modify the growth rate.

The quality of light directly affects plant growth, for example, it triggers germination. Generally, however, light quality is not under the control of the one who grows plants. Except for special situations it is quite well established that the full spectrum of sunlight is generally most satisfactory for plant growth.

Photoperiodism

Photoperiodism is a term describing the behavior of a plant in relation to day length. Based upon their reaction to day length, plants are classified as short-day, long-day or indeterminate plants. Short-day plants flower only under short-day conditions. If they are grown under long-day photoperiods they will not flower and continue to grow vegetatively. Examples of short-day plants are most of the spring flowers and such autumn-flowering plants as ragweed, asters and scarlet sage.

Long-day plants attain their flowering stage only when the length of day falls within certain limits, usually 12 hours or longer. Such plants are the radish, lettuce, grains, clover and many others that normally bloom in midsummer.

Still other plants such as the tomato, cotton and buckwheat flower complete their reproductive cycles over a wide range of day lengths. These plants are termed indeterminate.

Length of day has also been reported to have an influence on the formation of tubers and bulbs, the character and extent of branching, root growth, abscission or the dropping of leaves, flowers and other plant parts, dormancy and other effects.

Photoperiodism plays a large part in whether varieties and species may be adaptable to other areas such as moving a plant species from the north to the south and vice versa. Artificial systems can be set up to induce plants to respond contrary to what they would do if left to grow in the open. The poinsettia is forced to change its growth pattern to produce the colorful display at Christmas time by altering its photoperiod.

In reality, the length of the dark period is the controlling factor in photoperiodicity, not the day length. This has been established with experimentation; however, since in normal situations long days have short nights and vice versa, it is natural to assign day length to photoperiodism.

Water and Growth

The importance of water to plant growth is readily apparent. Water is required in photosynthesis, is a part of protoplasm and serves as a vehicle for translocation of food and mineral elements. The availability of water may influence the form, structure and nature of plant growth. Plants absorb more water than any other soil constituent. As previously indicated, a large part of the moisture taken up is transpired. Since water is such an important factor in plant growth and overall crop production, a separate chapter, Chapter 2, is devoted to this subject.

The Atmosphere and Growth

Quantities of nitrogen, oxygen and carbon dioxide do not vary except locally in the atmosphere. Air normally contains 78 percent nitrogen, 21 percent oxygen and 0.03 percent carbon dioxide. No higher plant is known which can make direct use of elemental nitrogen except through the action of certain microorganisms. All plants need oxygen for respiration, and the atmosphere is the most common source. Some plants are able to utilize the oxygen from oxidized compounds such as nitrates and sulfates but this is normally confined to the microorganisms or specialized plants. All photosynthesizing plants require carbon dioxide, and this becomes the basic process by which carbon is fixed into organic matter.

The atmosphere often contains gases, particulate matter and other contaminants which can have a direct effect on plant growth. Some of these effects are positive. For example, sulfur dioxide at low levels can be absorbed by the aerial portions of some plants, and much of the nutrient need for sulfur can be satisfied this way.

Specific damage to growing plants has been observed from the pollutants in air. Such products as ozone, PAN (peroxyacetyl nitrates), excess sulfur dioxide, ethylene, fluorides and others come from motor vehicles, combustion of fuels, organic solvents and other sources. Most damage has been noted on fruit and vegetable crops, pine trees and flowers. Usually the injury shows up on leaves, but sometimes the plants are stunted or produce poorly.

Mineral Nutrient Requirements and Growth

Present information indicates the need for 16 elements in plant growth. These are carbon, hydrogen, oxygen, nitrogen, phosphorus, potassium, calcium, magnesium, sulfur, boron, chlorine, copper, iron, manganese, molybdenum and zinc. Many more elements are found in plants but their essentiality has not been established. Some of these, listed alphabetically, are as follows: aluminum, arsenic, barium, bromine, cobalt, fluorine, iodine, lithium, nickel, selenium, silicon, sodium, strontium, titanium and vanadium. This is not a complete list since practically all of the known elements have been isolated at one time or another from plant materials. Functions of the nutrient elements in the plant, their supply in the soil and additions in fertilizers are discussed in other chapters.

By necessity, many of the factors associated with plant growth
have been omitted from this short chapter. The discussion has been
simplified, and much of the basic biology, chemistry and related
scientific disciplines has only been touched upon. It was written to
help the student, the farmer or anyone interested to understand
something about plant growth.

If one desires to aquaint himself more completely, he is referred
to the references at the end of this chapter.

SUPPLEMENTARY READING

1. *Botony—Brief Introduction to Plant Biology.* T. L. Rost, M. G. Barbour,
   R. M. Thornton, T. E. Weir and C. R. Stocking. John Wiley & Sons, Inc.
   1979.
2. *Modern Plant Biology.* Howard J. Dittmer. Van Nostrand Reinhold Company. 1972.
3. *Physiology of Plant Growth and Development.* M. B. Wilkins. McGraw-Hill
   Book Company. 1969.
4. *Plant Growth.* Michael Black and Jack Edelman. Harvard University Press.
   1970.
5. *Plant Physiology*, Second Edition. F. B. Salisbury and C. W. Ross. Wads-
   worth Publishing Co. 1979.
6. *Soil Conditions and Plant Growth*, Tenth Edition. E. J. Russell. Halstead
   Press. 1973.

Chapter 4

ESSENTIAL PLANT NUTRIENTS

There are more than 100 chemical elements known to man today. Only 16 of these have been shown to be essential to plants. Others may be found to be essential in the future. A few have already demonstrated an ability to stimulate plant growth under certain conditions.

Three of the 16 essential elements, carbon, hydrogen and oxygen, are taken primarily from the air and water. The other 13 essential elements are normally absorbed from soil by plant roots. These 13 elements are divided into three groups, primary nutrients, secondary nutrients, and micronutrients. This grouping separates the elements on the basis of relative amounts required for plant growth. None of these elements is any more essential than the others, regardless of amounts required.

CARBON, HYDROGEN AND OXYGEN

Carbon forms the skeleton for all organic molecules. Hence it is a basic building block for plant life. Carbon is taken from the atmosphere by plants in the form of carbon dioxide. Through the process of photosynthesis, carbon is combined with hydrogen and oxygen to form carbohydrates. Further chemical combinations, some with other essential elements, produce the myriads of substances required for plant growth.

Oxygen is required for respiration in plant cells whereby energy is derived from the breakdown of carbohydrates. Many compounds required for plant growth processes contain oxygen. Hydrogen along with oxygen forms water, which constitutes a large proportion of the total weight of plants. Water is required for transport of minerals and plant food and it also enters into many chemical reactions necessary for plant growth. Hydrogen is also a constituent of many other compounds necessary for plant growth. Since carbon, hydrogen and oxygen are supplied to plants primarily from the air and from water, the concern over their supply is somewhat different from that of the other 13 essential elements.

PRIMARY PLANT NUTRIENTS

Nitrogen

Nitrogen is taken up by plants primarily as nitrate (NO_3^-) or ammonium (NH_4^+) ions. Plants can utilize both of these forms of nitrogen in their growth processes.

Most of the nitrogen taken up by plants is in the nitrate form. There are two basic reasons for this. First, nitrate nitrogen is mobile in the soil and moves with soil water to plant roots where uptake can occur. Ammonic nitrogen, on the other hand, is bound to the surfaces of soil particles and cannot move to the roots. Second, all forms of nitrogen added to soils are changed to nitrate under proper conditions of temperature, aeration, moisture, etc., by soil organisms.

Nitrogen is utilized by plants to synthesize amino acids which in turn form proteins. The protoplasm of all living cells contains protein. Nitrogen is also required by plants for other vital compounds such as chlorophyll, nucleic acids and enzymes.

Soil Nitrogen

Most of the nitrogen in soils is unavailable to growing crops because it is tied up in organic matter. Only about 2 percent of this nitrogen is made available to crops each year. Since western soils generally contain relatively small amounts of organic matter, the amount of nitrogen made available for crop use each year is not great, perhaps about 20 pounds per acre.

Many reactions involving nitrogen occur in the soil. Most of them are the result of microbial activity. Nitrogen is made available to crops from organic matter through two of these reactions. Protein and allied compounds are broken down into amino acids through a reaction called *aminization*. Soil organisms acquire energy from this digestion. They also utilize some of the amino nitrogen in their own cell structure. Ammonic nitrogen is formed by the second reaction, which converts amino compounds into ammonia (NH_3) and ammonium (NH_4^+) compounds. This reaction is called *ammonification*. The two reactions, aminization and ammonification, are referred to as *mineralization*.

Ammonic forms of nitrogen are changed to nitrate by two distinct groups of microbacteria. *Nitrosomonas* and *Nitrosococcus* convert ammonia to nitrite:

$$2NH_4^+ + 3O_2 \longrightarrow 2NO_2^- + 2H_2O + 4H^+ + \text{energy}$$

ammonium   oxygen   nitrite   water   hydrogen
nitrogen   ions

Nitrobacter oxidizes nitrite to nitrate:

$$2NO_2^- + O_2 \longrightarrow 2NO_3^- + energy$$

nitrite   oxygen   nitrate

This two-step reaction is called *nitrification*. The reactions occur readily under conditions of warm temperature, adequate oxygen and moisture and optimum pH.

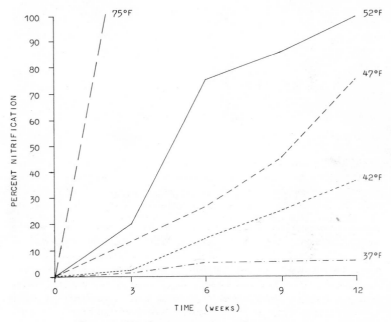

Fig. 4-1. Nitrification at various soil temperatures.

At 75°F, nitrification may be completed in one to two weeks (Figure 4-1), at 50°F, 12 weeks or more may be required. Figure 4-2 shows the relation of soil pH to nitrification. For optimum conversion, pH must be maintained between 5.5 and 7.8.

Nitrogen may be lost from the soil to the atmosphere by reactions that convert nitrate to gaseous compounds of nitrogen. This process is called *denitrification*. Under anaerobic conditions caused by excessive moisture and/or soil compaction, certain bacteria are capa-

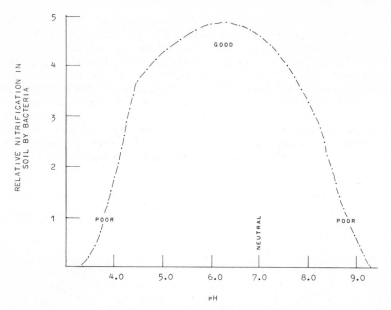

Fig. 4-2.   Relation of soil pH to nitrification.

ble of removing oxygen from chemical compounds in the soil to meet the needs of their life processes. When nitrate is used, various gases such as nitrous oxide (N_2O, nitric oxide (NO) and nitrogen (N_2) are formed. The reaction can be represented as follows:

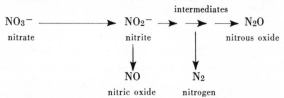

As these gases are lost from the soil into the atmosphere, there is a loss of crop-producing nitrogen from the soil.

The Nitrogen Cycle

The atmosphere contains approximately 78 percent nitrogen. It is estimated that over every acre of land there are some 35,000 tons of nitrogen. In order for crops to utilize this nitrogen, it must be combined with hydrogen or oxygen. This process is called *nitrogen fixation*. Nitrogen may be fixed by various soil organisms. Some of these live in nodules on roots of legumes, and others are free living organisms. Lightning also fixes smaller amounts of nitrogen which are carried into the soil by rain. The fertilizer industry fixes several million tons of nitrogen each year in various nitrogen fertilizers.

Figure 4-3 shows that nitrogen fixation by these various means provides nitrogen for growing crops. Crops in turn provide nitrogen for animals which use the crops for food. Plant and animal wastes are returned to the soil, carrying nitrogen with them.

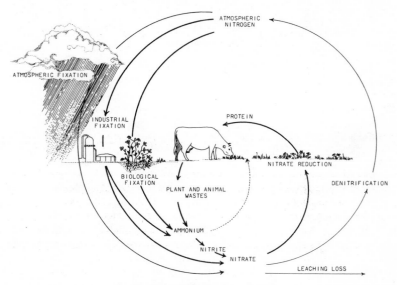

Fig. 4-3. The nitrogen cycle.

Nitrogen may be lost to the atmosphere by denitrification or by volatilization where ammonia is produced at or near the soil surface. Leaching of nitrates, primarily from light soils may move nitrogen below the root zone where it cannot be utilized by crops. Erosion of surface soil may also carry nitrogen from fields into streams and lakes or to the oceans. This continuous recycling of nitrogen is called the *nitrogen cycle.*

Symptoms of nitrogen deficiency in plants:
1. Slow growth; stunted plants.
2. Yellow-green color (chlorosis).
3. "Firing" of tips and margins of leaves beginning with more mature leaves.

Chlorosis is usually more pronounced in older tissue since nitrogen is mobile within plants and tends to move from older to younger tissue when nitrogen is in short supply.

Phosphorus

Phosphorus is absorbed by plants as $H_2PO_4^-$, $HPO_4^=$ or $PO_4^=$, depending upon soil pH. Most of the total soil phosphorus is tied up chemically in compounds of limited solubility. In neutral to alkaline soils, calcium phosphate is formed, while in acid soils, iron and aluminum phosphates are produced.

Available soil phosphorus may be only 1 percent or less of the total amount present. Solubility of phosphate is controlled by several factors. The total amount of precipitated phosphate in the soil is one factor. The greater the total amount present in the soil, the better the chance of having more phosphorus in solution. Another important factor is the extent of contact between precipitated phosphate and the soil solution. Greater exposure of phosphate to soil solution and to plant roots increases the ability to maintain replacement supplies. During periods of rapid growth, phosphorus in the soil solution may be replaced 10 times or more per day from solid phase phosphorus. Soil temperature and pH also affect the solubility of phosphate. Maximum availability of soil phosphorus occurs at pH 6.5 to 7.0.

Research by several workers has demonstrated an increased uptake of phosphorus by plants when nitrogen is added with the phosphate fertilizer. This synergistic effect is especially true for banded phosphorus applications. Various explanations have been proposed for this observation. Increased root growth, physiological changes making the root cells more receptive to phosphorus, increased transfer of phosphorus across the root to the xylem and a lowering of soil pH from ammonium nitrogen have all been suggested as reasons for the increased efficiency of phosphorus uptake in the presence of nitrogen.

Phosphorus is present in all living cells. It is utilized by plants to form nucleic acids (DNA and RNA). It is used in storage and transfer of energy through energy-rich linkages (ATP and ADP).

Phosphorus stimulates early growth and root formation. It hastens maturity and promotes seed production. Phosphorus supplementation is required most by crops under these circumstances: (1) growth in cold weather, (2) limited root growth and (3) fast top growth. Lettuce is an example of a very highly responsive crop due to the conditions outlined. Legumes such as alfalfa and beans are large users of phosphoric fertilizers. Least responsive crops are trees and vines with extensive root systems, long growing seasons and production during warmer times of the year.

Symptoms of phosphorus deficiency in plants:

1. Slow growth; stunted plants.
2. Purplish coloration on foliage of some plants.
3. Dark green coloration with tips of leaves dying.
4. Delayed maturity.
5. Poor grain, fruit or seed development.

Potassium

Potassium is taken up by plants in the form of potassium ions (K^+). It is not synthesized into compounds such as occurs with nitrogen and phosphorus, but tends to remain in ionic form within cells and tissues. Potassium is essential for translocation of sugars and for starch formation. It is required in the opening and closing of stomata by guard cells. Potassium encourages root growth and increases crop resistance to disease. It produces larger, more uniformly distributed xylem vessels throughout the root system. Potassium increases size and quality of fruit and grains and is essential for high quality forage crops.

Soils may contain 40,000 to 60,000 pounds of potassium per acre. About 90 to 98 percent of the potassium occurs in primary minerals and is unavailable to crops. One to 10 percent is trapped in expanding lattice clays and is only slowly available. One to 2 percent is contained in the soil solution and on exchange sites and is readily available to crops.

Potassium has been found to be most required by tree crops such as prunes with a very high carbohydrate production. The most responsive vegetable has been the potato, which again is a high producer of carbohydrate as starch in the tubers. Potassium is removed from the soil in larger amounts where the entire vegetative growth is removed such as in high yields of silage, hays and celery (see Table 4-1). Potassium is mobile in plant tissues.

Symptoms of potassium deficiency in plants:

1. Tip and marginal "burn" starting on more mature leaves.
2. Weak stalks, plants "lodge" easily.

Table 4-1.   Plant Food Utilization by Various Crops*

		Pounds per Acre		
Crop	Yield	N	P_2O_5	K_2O
Field crops				
Barley	2½ t.	160	60	160
Corn (grain)	5 t.	240	100	240
Corn (silage)	30 t.	250	105	250
Cotton (lint)	1,500 lbs.	180	65	125
Grain sorghum	4 t.	250	90	200
Oats	3,200 lbs.	115	40	145
Rice	7,000 lbs.	110	60	150
Safflower	4,000 lbs.	200	50	150
Soybeans	3,600 lbs.	325	65	145
Sugar beets	30 t.	255	60	550
Wheat	3 t.	175	70	200
Vegetable crops				
Asparagus	3,000 lbs.	95	50	120
Beans (snap)	10,000 lbs.	175	40	200
Broccoli	18,000 lbs.	80	30	75
Cabbage	35 t.	270	65	250
Celery	75 t.	280	165	750
Lettuce	20 t.	95	30	200
Potatoes (Irish)	500 cwt.	270	100	550
Squash	10 t.	85	20	120
Sweet potatoes	15 t.	155	70	315
Tomatoes	30 t.	180	50	340
Fruit and nut crops				
Almonds (in shell)	3,000 lbs.	200	75	250
Apples	15 t.	120	55	215
Cantaloupes	30 t.	220	70	400
Grapes	15 t.	125	45	195
Oranges	30 t.	265	55	330
Peaches	15 t.	95	40	120
Pears	15 t.	85	25	95
Prunes	15 t.	90	30	130
Forage crops				
Alfalfa	8 t.	480	95	480
Bromegrass	5 t.	220	65	315
Clovergrass	6 t.	300	90	360
Orchardgrass	6 t.	300	100	375
Sorghum-sudan	8 t.	325	125	475
Timothy	4 t.	150	55	250
Vetch	7 t.	390	105	320
Turf crops				
Bentgrass	2½ t.	260	65	145
Bermudagrass	4 t.	225	40	160

*Total uptake in harvested portion.

3. Small fruit or shriveled seeds.
4. Slow growth.

SECONDARY PLANT NUTRIENTS

Calcium

Calcium is absorbed by plants as the calcium ion (Ca^{++}). It is an essential part of cell wall structure and must be present for the formation of new cells. It is believed to counteract toxic effects of oxalic acid by forming calcium oxalate in the vacuoles of cells. Calcium is non-mobile in plants. Young tissue is affected first under conditions of deficiency.

Calcium has only been required as a foliar spray for certain celery varieties to prevent a disorder of the stalk called "brown checking." Calcium is so generally abundant that its only other requirement as a fertilizer nutrient has been on very acid soils where lime is required.

Symptoms of calcium deficiency in plants:
1. Death of growing points (terminal buds) on plants. Root tips also affected.
2. Abnormal dark green appearance of foliage.
3. Premature shedding of blossoms and buds.
4. Weakened stems.

Magnesium

Plant uptake of magnesium is in the form of the magnesium ion (Mg^{++}). The chlorophyll molecule contains magnesium. It is therefore essential for photosynthesis. Magnesium serves as an activator for many plant enzymes required in growth processes.

Magnesium is mobile within plants and can be readily translocated from older to younger tissue under conditions of deficiency.

Magnesium has generally high levels of availability in western soils, but is more often deficient than calcium. The most common requirements for magnesium have been on celery and citrus. These crops are likely in need of magnesium to balance the generally high use of potassium from fertilizers and manure. Crops growing in sandy soils may also show deficiencies.

Symptoms of magnesium deficiency in plants:
1. Interveinal chlorosis (yellowing) in older leaves.
2. Curling of leaves upward along margins.
3. Marginal yellowing with green "Christmas tree" area along mid-rib of leaf.

Sulfur

Uptake of sulfur by plants is in the form of sulfate ions ($SO_4^=$). Sulfur may also be absorbed from the air through leaves in areas where the atmosphere has been enriched with sulfur compounds from industrial sources.

Sulfur is a constituent of three amino acids (cystine, methionine and cysteine) and is therefore essential for protein synthesis. It is essential for nodule formation on legume roots. Sulfur is present in oil compounds responsible for the characteristic odors of plants such as garlic and onion.

Sulfur fertilization is required most by legume crops. The geographical distribution of sulfur deficiencies in California is rather well identified as the eastern foothills in the San Joaquin Valley, the central coastal range and the Sacramento Valley with its contributing valleys. Sulfur is generally deficient in Oregon, Washington and Idaho. In these states the sulfur-supplying power of the soil is frequently so low that even grain crops require sulfur fertilization.

Symptoms of sulfur deficiency in plants:
1. Young leaves light green to yellowish in color. In some plants, older tissue may be affected also.
2. Small and spindly plants.
3. Retarded growth rate and delayed maturity.

MICRONUTRIENTS

Even though micronutrients are used by plants in very small amounts, they are just as essential for plant growth as the larger amounts of primary and secondary nutrients. Care must be exercised in the use of micronutrients, since the difference between deficient and toxic levels is small. Micronutrients should not be applied as a "shotgun" application to cover possible deficiencies. They should be applied only when the need has been demonstrated.

Zinc

Zinc is an essential constituent of several important enzyme systems in plants. It controls the synthesis of indoleacetic acid, an important plant growth regulator. Terminal growth areas are affected first when zinc is deficient. Zinc is absorbed by plants as the zinc ion (Zn^{++}).

Zinc has been the micronutrient most often needed by western crops. Following nitrogen and phosphorus it is the nutrient most frequently used. Citrus generally is given zinc as part of a foliar spray program one to several times a year. Many other tree crops, grapes, beans, onions, tomatoes, cotton, rice and corn have generally required zinc fertilization. The great increase in corn acreage has brought considerable attention to the need for zinc.

Symptoms of zinc deficiency in plants:

1. Decrease in stem length and a rosetteing of terminal leaves.
2. Reduced fruit bud formation.
3. Mottled leaves (interveinal chlorosis).
4. Dieback of twigs after first year.
5. Striping or banding on corn leaves.

Iron

Iron is required for the formation of chlorophyll in plant cells. It serves as an activator for biochemical processes such as respiration, photosynthesis and symbiotic nitrogen fixation. Iron deficiency can be induced by high levels of manganese or high lime content in soils. Iron is taken up by plants either as ferrous (Fe^{++}) or ferric (Fe^{+++}) ions.

Iron is usually contained in ample amounts in western soils. Where deficiencies do occur, it is likely due to an imbalance (excess zinc or manganese), high pH or poor aeration. Crops most often affected are grasses such as sorghum and turf, certain tree crops and ornamentals.

Symptoms of iron deficiency in plants:

1. Interveinal chlorosis of young leaves. Veins remain green except in severe cases.
2. Twig dieback.
3. In severe cases, death of entire limbs or plants.

Manganese

Manganese serves as an activator for enzymes in growth processes. It assists iron in chlorophyll formation. High manganese concentration may induce iron deficiency. Manganese uptake is primarily in the form of the ion (Mn^{++}).

Manganese is generally required with zinc in foliar spraying of citrus. Other tree crops may show deficiencies, but otherwise, there is no common recognition of requirements for this element in fertilizer programs.

Symptoms of manganese deficiency in plants:

1. Interveinal chlorosis of young leaves. Gradation of pale

green coloration with darker color next to veins. No sharp
distinction between veins and interveinal areas as with iron
deficiency.
2. Development of gray specks (oats), interveinal white
   streaks (wheat), or interveinal brown spots and streaks
   (barley).

Copper

Copper is an activator of several enzymes in plants. It may play
a role in vitamin A production. A deficiency interferes with protein
synthesis. Plant uptake is in the form of ions (Cu^+, Cu^{++}).

Native copper supply has been recognized only rarely as needing
supplementation. These few instances have been with tree crops and
some crops on organic soils and sands. Copper is rather highly toxic
at low levels and is therefore not recommended except where the
need has been established.

Symptoms of copper deficiency in plants:
1. Stunted growth.
2. Dieback of terminal shoots in trees.
3. Poor pigmentation.
4. Wilting and eventual death of leaf tips.
5. Formation of gum pockets around central pith in oranges.

Boron

Boron is taken up by plants as the borate ion. It functions in plants
in differentiation of meristem cells. With boron deficiency, cells may
continue to divide, but structural components are not differentiated.
Boron also apparently regulates metabolism of carbohydrates in
plants. Boron is non-mobile in plants, and a continuous supply is
necessary at all growing points. Deficiency is first found in the
youngest tissues of the plant.

Boron has been as often toxic in western soils as it has been
deficient. The toxicities occur most often in inland desert areas
associated with high boron waters. Deficiencies are generally related to high rainfall areas as well as areas irrigated for a considerable time with low boron surface waters. Most boron fertilization
has been required by tree crops and legumes, in central and northern California and in Oregon and Washington.

Symptoms of boron deficiency in plants:
1. Death of terminal growth, causing lateral buds to develop
   and producing a "witches broom" effect.

Plate 4-1. Nitrogen deficiency on corn. Lower leaves show characteristic yellowing (chlorosis), while upper portion of plants remain green.

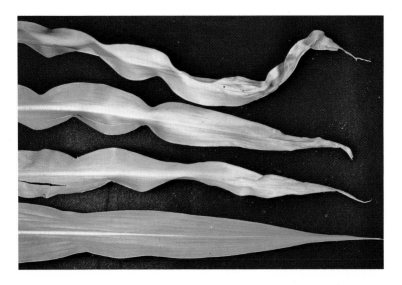

Plate 4-2. Nitrogen deficiency on corn leaves. Lower leaf is normal, upper leaves show characteristic yellowing and "firing" of tissue. Yellowing starts at tips of leaves and moves toward the base in a "V" pattern down the mid-rib.

Plate 4-3. Phosphorus deficiency on corn. Stunted growth and purple coloration are characteristic symptoms on many plants. Symptoms usually appear during early growth while soils are cold and root systems are small.

Plate 4-4. Potassium deficiency on alfalfa. Yellow spots develop on margins of more mature leaves. Specks may coalesce to produce entire yellow margins on leaves.

Plate 4-5. Potassium deficiency on cotton. Mature leaves show marginal and interveinal chlorosis. Marginal tissue may die as deficiency progresses.

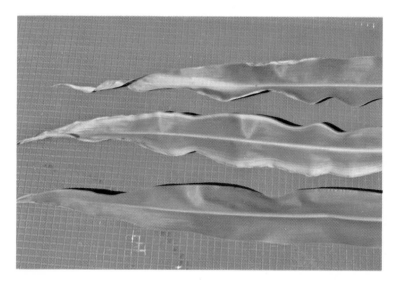

Plate 4-6. Potassium- and nitrogen-deficient leaves on grain sorghum. Center leaves show characteristic potassium deficiency symptoms with chlorotic (yellow) and necrotic (dead) tissue along margins of more mature leaves. The upper leaf shows typical yellow "V" progressing from tip toward base of leaf. Bottom leaf is normal (for comparison).

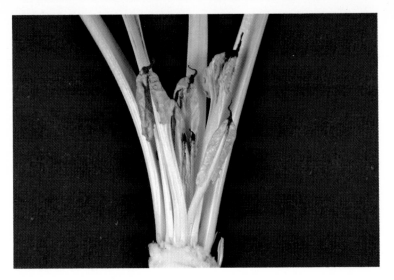

Plate 4-7. Calcium deficiency on sugar beets. Calcium is non-mobile in the plant, and young tissue is affected first. Dead tissue on tips of new leaves is a very characteristic symptom of calcium deficiency on sugar beets.

Plate 4-8. Calcium deficiency on sugar beets. Growing tips are affected first with calcium deficiency. Death of tap root has caused "sprangle" root appearance. Center beet is normal (for comparison).

Plate 4-9. Magnesium deficiency on citrus (orange). Chlorosis of tip and margins of more mature leaves produces a "Christmas tree" pattern along midrib of leaf.

Plate 4-10. Boron deficiency on almond. Death of terminal growth causes lateral buds to develop. A characteristic "witches broom" appearance occurs on many small lateral branches.

Plate 4-11. Iron deficiency on pears. Yellowing of younger leaves. Severe deficiency on center tree.

Plate 4-12. Iron and manganese deficiencies on citrus (lemon). Two center leaves show typical iron deficiency symptoms: green veins with yellowing of interveinal areas of young leaves. The two outer leaves show characteristic manganese deficiency. Younger leaves show interveinal chlorosis with gradation of pale green coloration, with darker color next to the veins.

Plate 4-13. Zinc deficiency on citrus (orange). Small narrow leaves with yellow mottling between the veins. It appears on younger leaves first.

Plate 4-14. Zinc deficiency on citrus (orange). Normal leaf (right); leaves with varying zinc deficiency symptoms to left. Small, narrow leaves with interveinal mottling are characteristic of zinc deficiency.

Plate 4-15. Boron excess on walnuts. Necrotic spots develop along leaf margins and into interveinal tissue.

Plate 4-16. Herbicide (Triazine) damage on almonds. Yellowing of leaf margins and interveinal tissue occurs on leaves of all stages of maturity. Veins and some tissue bordering veins remain green.

2. Thickened, curled, wilted and chlorotic leaves.
3. Soft or necrotic spots in fruit or tubers.
4. Reduced flowering or improper pollination.

Molybdenum

Molybdenum is required by plants for utilization of nitrogen. Plants cannot transform nitrate nitrogen into amino acids without molybdenum. Legumes cannot fix atmospheric nitrogen symbiotically unless molybdenum is present.

Molybdenum has been found in quantities toxic to livestock due to high concentrations in forage grown in inland desert areas, such as the San Joaquin Valley and Nevada. As soils from these areas are farmed more, the problem has been reduced. Deficiencies of molybdenum have required small additions (ounces per acre) to legumes growing in eastern Washington, Oregon and in parts of Idaho.

Symptoms of molybdenum deficiency in plants:
1. Stunting and lack of vigor. This is similar to nitrogen deficiency due to the key role of molybdenum in nitrogen utilization by plants.
2. Marginal scorching and cupping or rolling of leaves.
3. "Whiptail" of cauliflower.
4. Yellow spotting of citrus.

Chlorine

Chlorine is required in photosynthetic reactions in plants. Deficiency is not seen in the field due to its universal presence in nature.

Symptoms of chlorine deficiency in plants:
1. Wilting, followed by chlorosis.
2. Excessive branching of lateral roots.
3. Bronzing of leaves.
4. Chlorosis and necrosis in tomatoes and barley.

NUTRIENT BALANCE

Balance is important in plant nutrition. An excess of one nutrient can cause reduced uptake of another. An excess of potassium, for example, may compete with magnesium uptake by crops. A heavy application of phosphorus may induce a zinc deficiency in soil that is marginal or low in zinc. Excess iron may induce a manganese deficiency.

Maintaining a balance of nutrients in the soil is an important management objective. By judicious use of fertilizers, nutrients which are deficient in soil can be applied to growing crops. The objective of fertilizer programs is to supplement the capacity of soils to supply adequate nutrients to growing crops.

DIAGNOSING NUTRIENT NEEDS

This topic is covered more completely in Chapter 9. No attempt will be made to discuss this topic other than to point out one or two precautions. Visual symptoms of nutrient deficiencies can be a useful tool for diagnosing problems. Assistance should be obtained from a qualified person, since chlorosis (yellowing) and necrosis (death) of tissues can result from problems other than nutrient deficiencies. Toxicity from excessive amounts of certain elements or damage from herbicides, as well as lack of proper aeration in the root zone, can produce yellowing or death of tissue. A trained person can usually distinguish between these various problems where the average person may have difficulty. Soil and/or tissue analyses should be used to verify nutrient deficiencies and to determine their causes prior to initiating a program for correction.

SUPPLEMENTARY READING

1. *Better Crops with Plant Foods.* The Potash and Phosphate Institute. 63: 4-7. Spring 1979.
2. *Diagnostic Criteria for Plants and Soils.* H. D. Chapman, Editor. University of California, Division of Agricultural Sciences. 1966.
3. *Hunger Signs in Crops,* Third Edition. H. B. Sprague, Editor. David McKay Co., Inc. 1964.
4. "Micronutrient Report." Kent B. Tyler. *Agribusiness News,* Desert Edition, 8:3. April 1972.
5. *Soil Fertility and Fertilizers,* Third Edition. S. L. Tisdale and W. L. Nelson. The Macmillan Company. 1975.

Chapter 5

FERTILIZERS—A SOURCE OF
PLANT NUTRIENTS

Soil serves as a storehouse for plant nutrients and normally provides a substantial amount of the crop's nutrient requirements. Under most conditions, however, crop production can be enhanced by proper application of supplemental nutrients. Any material containing one or more of the essential nutrients that are added to the soil or applied to plant foliage for the purpose of supplementing the plant nutrient supply can be called a fertilizer.

The earliest fertilizer materials were animal manures, plant and animal residues, ground bones and potash salts derived from wood ashes. Three major developments in the nineteenth century were the forerunners of the modern fertilizer industry:

1839—The discovery of potassium salt deposits in the German states.

1842—The treatment of ground phosphate rock with sulfuric acid to form superphosphate.

1884—The development of the theoretical principles for combining hydrogen and atmospheric nitrogen to form ammonia.

TYPES OF FERTILIZERS

Based upon their primary nutrient content (N, P_2O_5, K_2O), fertilizers are designated as single nutrient or multinutrient fertilizers. Single nutrient fertilizers are called "materials" or "simple" fertilizers. Multinutrient fertilizers are referred to as "mixed fertilizers" or "complexes."

Multinutrient fertilizers are given a numerical designation consisting of three numbers. These numbers represent respectively the nitrogen (N), phosphate (P_2O_5) and potash (K_2O) content of the fertilizer in terms of percent by weight. This three-number designation is called a "grade."

The fertilizer "ratio" is the relative proportion of each of the primary nutrients. For example, a 12-12-12 grade is a 1-1-1 ratio, and a 21-7-14 grade is a 3-1-2 ratio.

A zero in a grade, or ratio designation, indicates that that particular nutrient is not included in the fertilizer. For example, the grade designation for ammonium nitrate is 34-0-0, and the ratio designation for a 20-10-0 grade is 2-1-0.

NITROGEN FERTILIZERS

Anhydrous Ammonia

Nitrogen from the atmosphere is the primary source of all nitrogen used by plants. This inert gas comprises about 78 percent of the earth's atmosphere. Chemical fixation of atmospheric nitrogen is largely in the form of ammonia, which is the principal source of most nitrogen fertilizers. The production of ammonia involves the reaction of nitrogen and hydrogen in the presence of a catalyst at a temperature that ranges from 400° to 500° C. The pressure required varies from 4,000 to 5,000 psig in the older type reciprocating compressor plants to pressures as low as 2,200 psig in the newer centrifugal compressor plants.

The reaction is expressed by the following equation:

$$3H_2 + N_2 \rightleftarrows 2NH_3$$

Properties of Anhydrous Ammonia

Color	Colorless
Odor	Pungent, sharp
Molecular weight	17
Pounds per gal. at 60° F	5.14
Pounds per cu ft. at 60° F	38.45
Boiling point at 1 atmosphere pressure	−28.03° F
Freezing point at 1 atmosphere pressure	−107.86° F
Calcium carbonate equivalent (lbs./100 lbs.)	148
Nitrogen content	82%

Gaseous ammonia is lighter than air. When compressed, it becomes a liquid about 60 percent as heavy as water. It is readily absorbed in water up to concentrations of 30 to 40 percent by weight.

The high vapor pressure of anhydrous ammonia at ordinary temperatures requires that it be transported in pressure containers, generally with a minimum working pressure of 265 psig.

Although the supply of nitrogen from the air is virtually limitless, sources of hydrogen are limited. In the United States, virtually all ammonia facilities use natural gas as a source of hydrogen. A ton of ammonia requires about 33,000 cubic feet of natural gas for its hydrogen requirement. Possible alternative sources of hydrogen are water, coal, oven gases and naphtha. Naphtha is frequently used in foreign plants.

Aqua Ammonia

Anhydrous ammonia dissolved in water forms aqua ammonia. As usually sold, aqua ammonia contains 20 percent nitrogen, which is present in the ammonium form. Although under normal temperature conditions, aqua ammonia has some free ammonia, the vapor pressure is low, making it possible to store this fertilizer in low-pressure tanks and to apply it with less expensive equipment than required for anhydrous ammonia.

To minimize loss of nitrogen, aqua (like anhydrous) ammonia should be injected below the soil's surface or below the surface of water when applied in a water run. Because of the much lower free ammonia, direct soil applications need not be made as deeply as anhydrous ammonia.

Ammonium Nitrate

Ammonium nitrate was not extensively used as a fertilizer until after World War II. It is manufactured by reacting nitric acid with anhydrous ammonia. Nitric acid is produced by the oxidation of NH_3 with air in the presence of a catalyst, usually platinum. The initial oxidation reactions are:

$$4NH_3 + 5O_2 \longrightarrow 4NO + 6H_2O$$

$$2NO + O_2 \longrightarrow 2NO_2$$

The NO_2 is then absorbed in water to give nitric acid.

$$3NO_2 + H_2O \longrightarrow 2HNO_3 + NO$$

Direct use of nitric acid is limited in the mixed fertilizer industry or for other agricultural purposes. The basic reaction for ammonium nitrate is:

$$HNO_3 + NH_3 \longrightarrow NH_4NO_3$$

The reacted material is concentrated and either prilled or spherodized. Ammonium nitrate is normally coated to prevent caking, and the commercial product carries 33.5 to 34.0 percent nitrogen.

Properties of Ammonium Nitrate

Color	White
Molecular weight	80
Pounds per cu ft.	45–62
Angle of repose	37°–40°
Critical relative humidity @ 68° F, 20° C	63.3
Calcium carbonate equivalent (lbs./100 lbs.)	60.0
Nitrogen content	33.5–34.0%

Ammonium Nitrate–Lime

This nitrogen fertilizer is widely used in Europe. It is produced by the addition of 40 percent powdered limestone or calcium carbonate to ammonium nitrate. The calcium carbonate is added to the concentrated ammonium nitrate solution prior to prilling or granulating. Most ammonium nitrate–lime contains 26 percent nitrogen.

Ammonium Sulfate

In the West, ammonium sulfate is a widely used fertilizer. It contains both nitrogen and sulfur. In fact, it contains more sulfur 24 percent) than nitrogen (21 percent). Ammonium sulfate is one of the oldest forms of solid nitrogen fertilizer.

Direct manufactured ammonium sulfate is made in a neutralizer-crystallizer unit by reacting anhydrous ammonia with sulfuric acid under vacuum or at atmospheric pressure.

The reaction is as follows:

$$2NH_3 + H_2SO_4 \longrightarrow (NH_4)_2SO_4$$

Crystallization is the key operation in the production of ammonium sulfate, involving two major steps: (1) the formation of nuclei in a supersaturated solution, and (2) the growth of these nuclei to the crystal size desired. Where ammonium sulfate is used in bulk blends, it is desirable to have the crystals approximately the same size as other components of the blend.

Properties of Ammonium Sulfate

Color	White
Molecular weight	132
Pounds per cu ft.	66-68

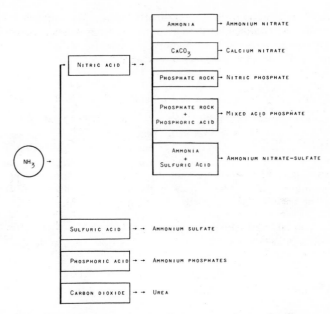

Fig. 5-1. Conversions of ammonia to various nitrogen fertilizers.

Angle of repose	28°
Critical relative humidity @ 68° F, 20° C	81
Calcium carbonate equivalent (lbs./100 lbs.)	110
Nitrogen content	21%
Sulfur content	24%

Ammonium Nitrate–Sulfate

Ammonium nitrate–sulfate is a relatively new dry nitrogen fertilizer in the U.S. This product has, however, been used in Europe for years.

Ammonium nitrate–sulfate is a double salt of ammonium nitrate and ammonium sulfate. It is manufactured by neutralizing a mixture of nitric and sulfuric acids with ammonia. The nitrogen content can be varied depending upon the relative percentages of ammonium nitrate and ammonium sulfate in the final product. One commercial product contains 30 percent nitrogen and 6.5 percent sulfur.

Ammonium nitrate–sulfate has good handling and storage properties and has proven to be an excellent fertilizer for crops, especially when the soil is deficient in both nitrogen and sulfur. Ammonium nitrate–sulfate is a very satisfactory product for direct application or for use in blended fertilizers.

Calcium Nitrate

Calcium nitrate is a white, hygroscopic material containing 15.5 percent nitrogen. It is produced by reacting nitric acid with crushed limestone (calcium carbonate). The mother liquor containing about 40 percent calcium nitrate is clarified before being brought to an 87 percent concentration in a vacuum evaporator. About 5 percent ammonium nitrate is added to bring the nitrogen content to 15.5 percent. The basic reaction for calcium nitrate production is:

$$CaCO_3 + 2HNO_3 \longrightarrow Ca(NO_3)_2 + CO_2 + H_2O$$

limestone   nitric   calcium   carbon   water
             acid   nitrate   dioxide

Calcium nitrate can also be produced by separation in the production of nitrophosphate fertilizers.

Nitrate of Soda

This nitrogen material contains 16 percent nitrogen. It is no longer manufactured in this country, and very small quantities of the natural product are being imported from Chile. Other products have largely replaced sodium nitrate because of its low concentration of nitrogen and its sodium content.

Urea

Urea is a widely used nitrogen fertilizer and is also used as a protein source in ruminant animal feeds. The raw materials for urea are ammonia and carbon dioxide. They are reacted together in a pressure vessel at temperatures between 170° and 210°C and pressures varying between 170 and 400 atmospheres (2,500–6,000 psig). The basic reactions are:

$$2NH_3 + CO_2 + H_2O \longrightarrow (NH_4)_2CO_3$$

ammonia   carbon   water   ammonium
          dioxide   carbonate

$$(NH_4)_2CO_3 \longrightarrow (NH_2)_2CO + 2H_2O$$

ammonium carbonate $\qquad\qquad$ urea $\quad$ water

The urea solution from these reactions has a concentration of about 80 percent. This solution may be used directly in urea solutions or it may be further concentrated to make dry urea. Fertilizer grade of dry urea contains about 46 percent nitrogen.

Properties of Urea

Color	White
Molecular weight	60
Pounds per cu ft.	46—48
Angle of repose	40°
Critical relative humidity @ 68° F, 20° C	80.7
Calcium carbonate equivalent (lbs./100 lbs.)	84
Nitrogen content	45—46%

Urea contains the highest percentage of nitrogen of commonly used solid fertilizers. Urea is very soluble and is less corrosive to equipment than many fertilizers. It is incompatible with some fertilizer materials, particularly those containing even small quantities of ammonium nitrate.

Nitrogen Solutions

Anhydrous ammonia and aqua ammonia have already been discussed. Because of the solubility of certain fertilizer salts, mainly ammonium nitrate and urea, these materials are widely used singularly or in combinations in aqueous solutions. For example, when ammonium nitrate and urea are mixed in equal proportions their solubilities are increased.

Many nitrogen solutions such as ammonium nitrate 20 percent solution, calcium ammonium nitrate 17 percent solution and urea-ammonium nitrate 32 percent solution are available (see Table 5-1). Nitrogen solutions may contain ammonium nitrate and urea nitrogen, the proportion of each being dependent on the product. They are classified as pressure and non-pressure solutions. The pressure solutions are those that have an appreciable vapor pressure because of the presence of more free ammonia than the solutions can hold.

Like any salt solution, nitrogen solutions will exhibit the phenomenon of salting out. Salting out is simply the precipitation of the dissolved salts when the temperature drops to a certain degree. This point is determined by the component salts of the solution.

Nitrogen solutions are readily stored if the appropriate storage

Table 5-1. Composition and Physical Properties of N Solutions

	Non-ammonia Solutions					Ammonia Solutions			Aqua Ammonia Solutions		
Total N %	20.0	20.0	28.0	30.0	32.0	37.0	37.0	41.0	20.0	20.6*	24.2
NH₃ %	–	–	–	–	–	16.6	15.8	22.2	24.4	25.0	29.4
NH₄NO₃ %	57.2	43.5	39.5	42.2	44.3	66.8	58.5	65.0	–	–	–
Urea %	–	–	30.5	32.7	35.4	–	7.7	–	–	–	–
Water %	42.8	56.5	30.0	25.1	20.3	16.6	18.0	12.8	75.6	75.0	70.6
Nitrate N %	10.0	–	7.0	7.4	7.8	11.7	10.2	11.4	–	–	–
Ammonia N %	10.0	–	7.0	7.4	7.8	25.3	23.2	29.6	20.0	20.6	24.2
Urea N %	–	20.0	14.0	15.2	16.4	–	3.6	–	–	–	–
Spec. grav. @ 60°F.	1.26	1.12	1.28	1.30	1.33	1.19	1.17	1.14	0.912	0.911	0.897
Lbs./gal. @ 60°F.	10.50	9.33	10.66	10.83	11.06	9.91	9.75	9.50	7.60	7.59	7.47
Lbs. N/gal. @ 60°F.	2.10	1.87	2.98	3.25	3.54	3.67	3.61	3.90	1.52	1.52	1.81
Vapor pressure psig @ 100°F.	–	–	–	–	–	1	2	10	1	2	10
Crystallization temp. °F.	41	52	1	15	32	56	28	21	–58	–103	–

*Generally sold at 20% N solution since some ammonia loss may occur in handling.

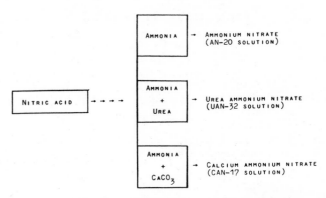

Fig. 5-2. Liquid nitrogen solutions.

materials are used. Table 5-2 describes corrosion rates of various metals, alloys and other materials.

Corrosion inhibitors in nitrogen solutions reduce corrosion of carbon (mild) steel. Ammonium thiocyanate (0.1 percent) and boron (0.1-0.4 percent) are commonly used inhibitors. A pH near 7.0 minimizes corrosion.

Table 5-2.   Corrosive Characteristics of Various Metals, Alloys and Other Materials by Non-pressure Nitrogen Solutions

Not Corroded	*Corroded*	*Materials Destroyed Rapidly*
Aluminum and aluminum alloys, types 3003, 3004, 5052, 5154 and 6061	Carbon steel Cast iron	Copper Brass Bronze
Stainless steel, types 303, 304, 316, 347 and 416		Monel Zinc
Rubber		Galvanized metals
Neoprene		Usual die castings
Polyethylene		Concrete
Vinyl resins		
Glass		

PHOSPHATE FERTILIZERS

Phosphate rock deposits are the basic source of all phosphate materials. The earth in certain areas is richly endowed with natural deposits of phosphate. These deposits may be of igneous or sedimentary origin, with the latter constituting the bulk of the world's deposits. The predominant phosphate mineral in most deposits is francolite, a carbonate fluorapatite represented by the formula $Ca_{10}F_2(PO_4)_6 \cdot CaCO_3$.

The principal world reserves now mined are in North Africa, North America and the USSR. The reserves in the United States are located principally in four areas: The western states of Montana, Idaho, Utah and Wyoming; in Florida; in North Carolina; and in Tennessee.

Prior to beneficiation, the phosphate matrix has a phosphate content of 14 to 35 percent. The phosphate must be concentrated for processing into fertilizers. The beneficiation process involves wet screening, hydroseparation and concentration by flotation. This product is then dried and may be calcined or ground prior to shipment. The P_2O_5 content is 31 to 33 percent.

There remains the task of converting the fluorapatite to more soluble forms that can be utilized by plants. A small quantity of finely ground rock phosphate is used on acid soils as a source of phosphorus. The pathways for treating rock phosphate used by the fertilizer industry and the end products produced are shown in Figure 5-3.

Phosphoric Acid and Superphosphoric Acid

Phosphoric acid is an important intermediate in the production of phosphatic fertilizers. There are two methods of producing phosphoric acid: the wet-process and the furnace-grade method. The wet-process method is the principal method used by the fertilizer industry. Practically all of the furnace acid is used for food and industrial purposes. It is more costly to produce.

Wet-process orthophosphoric acid is produced by the action of sulfuric acid on finely ground phosphate rock. The principal chemical reaction in simplified form is:

$$Ca_{10}F_2(PO_4)_6 + 10H_2SO_4 + 20H_2O \longrightarrow 10CaSO_4 \cdot 2H_2O + 2HF + 6H_3PO_4$$

| phosphate rock | sulfuric acid | water | gypsum | hydrogen fluoride | phosphoric acid |

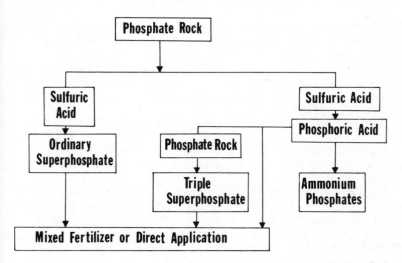

Fig. 5-3. Pathways for treating phosphate rock in the production of phosphatic fertilizers.

The acid is separated from the gypsum by filtration and washing to remove as much acid as possible from the gypsum cake. The acid produced contains about 30 percent P_2O_5 and is generally concentrated to the range of 40 to 54 percent P_2O_5.

Wet-process acid will frequently be green or black in color as a result of impurities including compounds of Fe, Al, Ca, Mg, F and organic matter. Some of these impurities may be beneficial as a source of micronutrients. A portion of the impurities may be present as solids.

Several by-products are produced during the manufacture of wet-process acid, but the principal one is impure gypsum. This by-product represents a disposal problem for many fertilizer manufacturers, but in areas of alkali soil, by-product gypsum is used as a soil amendment. A small amount of residual phosphorus remains in the gypsum.

The trend in the fertilizer industry is to make and use higher analysis compounds. This trend has given impetus to the development of superphosphoric acid, which is a condensation product of orthophosphoric acid. The condensation step is illustrated in Figure 5-4.

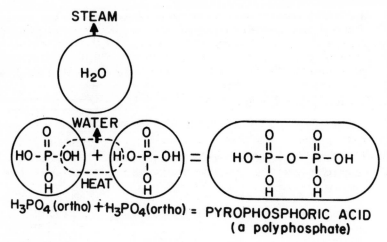

Fig. 5-4. Removal of water from orthophosphoric acid to produce pyrophosphoric acid.

The linking of two orthophosphoric acid molecules produces pyrophosphoric acid, three molecules gives tripoly, etc., as shown in Figure 5-5.

Collectively such an acid solution is called polyphosphoric acid or superphosphoric acid, and can be defined as any series of phos-

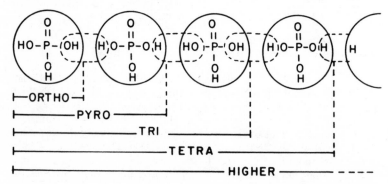

Fig. 5-5. The linkage of various members of orthophosphoric acid molecules to produce various polyphosphoric acids.

phoric acid whose molecular structure contains more than one atom of phosphorus such as pyrophosphoric acid ($H_4P_2O_7$), tripolyphosphoric acid ($H_5P_3O_{10}$), tetrapolyphosphoric acid ($H_6P_4O_{13}$), etc. The relationship between the concentration of acid and the type of acid species is shown in Table 5-3.

Normal Superphosphate

Normal superphosphate is produced by reacting sulfuric acid with finely ground phosphate rock. This is done by a batch or by a continuous process. The basic chemical reaction in simplified form is:

$$Ca_{10}F_2(PO_4)_6 + 7H_2SO_4 + 17H_2O \rightarrow 3Ca(H_2PO_4)_2 \cdot H_2O + 7CaSO_4 \cdot 2H_2O + 2HF$$

phosphate rock	sulfuric acid	water	monocalcium phosphate	gypsum	hydrogen fluoride

The highly insoluble phosphate in the phosphate rock is converted to monocalcium phosphate monohydrate where the phosphate is approximately 85 percent water soluble. The gypsum by-product from this reaction is intimately mixed with the monocalcium phosphate.

The sulfuric acid and phosphate rock are combined and mixed for one to two minutes. The resulting plastic mass is discharged into a compartment called a den. Retention time in the den varies from 1 to 24 hours. The acidulated phosphate sets up into a hard block and is removed from the den by various mechanical excavators. These excavators are equipped with knives to cut the block as it leaves the den. The material is stored for several weeks to allow completion of the chemical reaction. The cured product is excavated from the pile and pulverized and usually granulated before shipment. The end product usually contains 20 percent available P_2O_5.

Concentrated Superphosphate

Similar procedures are involved in the production of concentrated superphosphate (sometimes called triple superphosphate). The principal difference is that concentrated superphosphate is made with phosphoric acid as the acidulant, in contrast to sulfuric acid in normal superphosphate. Finely ground phosphate rock and phosphoric acid in the proper proportions are mixed together. The general reaction that occurs is:

$$Ca_{10}F_2(PO_4)_6 + 14H_3PO_4 + 10 H_2O \longrightarrow 10 Ca(H_2PO_4)_2 \cdot H_2O + 2HF$$

phosphate rock	phosphoric acid	water	monocalcium phosphate	hydrogen fluoride

Table 5-3. Forms of Phosphoric Acid at Various Concentrations*

Weight % P_2O_5	Equivalent % H_3PO_4	% of Total Phosphorus as				
		Ortho-phosphate	Pyro-phosphate	Tripoly-phosphate	Tetrapoly-phosphate	Higher Poly-phosphate
54	75	100				
68.8	95	100				
70	97	96	4			
72	99	90	10			
75.5	104	53	40	7		
77	106	40	47	11	2	
80	110	13	35	25	14	13
85	117	2	7	8	11	72

* These data are for furnace-grade acid. Wet-process acid data are generally different, and variation may occur between wet-process acid manufacturers.

The end product contains approximately 46 percent P_2O_5 and little $CaSO_4$.

Most superphosphates, used direct or in blends, are in granular form. Cured normal and concentrated superphosphate can be granulated through the addition of water and steam in a rotary drum granulator followed by drying and screening.

NITROGEN-PHOSPHATE COMBINATIONS

Ammonium Phosphates

The term *ammonium phosphate* encompasses a wide variety of fertilizers produced by ammoniation of phosphoric acid, often in a mixture with other materials. Such fertilizers may contain ammonium sulfate or ammonium nitrate as an additional source of nitrogen. The ammonium phosphate may be present as the diammonium or monoammonium salt, or a mixture of the two. The ammonium phosphates are widely used in bulk blends and for direct application. Commonly used ammonium phosphates are monoammonium phosphate 11-48-0, diammonium phosphate 16-48-0 and 18-46-0 and ammonium phosphate–sulfate 16-20-0. Liquid forms of ammonium phosphates commonly used are 8-24-0, 9-30-0 and 10-34-0.

Ammonium phosphates are produced by reacting ammonia with phosphoric acid in a preneutralization vessel and further ammoniating the slurry in a rotating ammoniation-granulation unit. Additional nitrogen, phosphate and sulfuric acid can be introduced into the ammoniation-granulation unit to make the desired grade. The material discharged from the unit is dried, screened and cooled before being placed in storage.

Solid ammonium polyphosphate fertilizers may be made by ammoniating superphosphoric acid. The acid is ammoniated in a water-cooled reactor at elevated temperature and pressure or in a pipe reactor at high temperature. The product is an anhydrous melt that is granulated by mixing it with cooled recycle material in a pugmill, followed by drying and cooling. The end analysis is 11-57-0.

Nitric Phosphates

There are three basic nitric phosphate processes but all utilize the same basic reaction:

$$Ca_{10}F_2(PO_4)_6 + 20\ HNO_3 \longrightarrow 6H_3PO_4 + 10\ Ca(NO_3)_2 + 2HF$$

| phosphate rock | nitric acid | | phosphoric acid | calcium nitrate | hydrogen fluoride |

If the resulting solution is simply ammoniated, dried and granulated, most of the phosphate is present as dicalcium phosphate. The modified processes used in the United States add phosphoric acid, which results in varying amounts of monoammonium phosphate in the final product. Modified nitric phosphates are perhaps better described as ammonium nitrate-phosphates.

POTASH FERTILIZERS

Potassium is found throughout the world in both soluble and insoluble forms. Today only the soluble forms are economically attractive. They occur primarily as chlorides and sulfates. Potassium chloride is by far the most important source of potash. Potassium sulfate and potassium nitrate are normally used where the chloride ion may result in poor crop quality or in a buildup of chloride in the soil. Table 5-4 gives the properties of the principal potassium salts.

Table 5-4. Properties of Potassium Salts

		Potassium Chloride	Potassium Sulfate	Potassium Nitrate
Color		white*	white	white
Molecular weight		74.5	174	101
Specific gravity		1.98	2.66	2.11
Melting point		772°C	1067°C	308°C
K_2O content, percent		63	54	46.6
Solubility in water (%K_2O)	0°C	13.6	3.7	5.5
	10°C	15.0	4.6	8.1
	20°C	16.1	5.4	11.2
	30°C	17.1	6.2	14.6
	40°C	18.1	7.0	18.2
	50°C	18.9	7.7	20.5
	60°C	19.8	8.6	24.2
	70°C	20.6	8.9	27.0
	80°C	21.3	9.5	29.2
	90°C	22.2	10.0	31.1
	100°C	22.9	10.5	33.2

* Impurities may impart a red color to some fertilizer grades.

The greatest proportion of potash produced today is from underground deposits. These underground deposits are mined by two principal means: (1) conventional shaft mining and (2) solution mining by pumping water underground to dissolve the ore. An important consideration of commercial feasibility of a deposit is its depth, which can vary from a few hundred feet to more than 4,000 feet below the surface. As yet there are no reports or drilling beyond 4,000 feet due to prohibitive costs. However, solution mining techniques could allow mining from depths below 4,000 feet.

The potash deposits in New Mexico vary from 500 to 2,500 feet below the surface. A deposit in Utah is located at about the 2,700-foot level. The Canadian beds in Saskatchewan tend to be closer to the surface in the north and go deeper as they approach the U.S. border. The northern beds now being worked by shaft mining techniques lie at 3,000 to 3,500 feet. At Regina, the beds are at about 5,000 feet and at the U.S. border, 7,000 feet below the surface.

An alternative source of potash is in natural brines in various

Fig. 5-6. Mining potash ore 900 feet below the earth's surface. Photo by Suter, Hedrich-Blessing.

parts of the world. Currently there are three natural brine deposits being worked in the world—Searles Lake, California, Salt Lake and Bonneville Lake in Utah and the Dead Sea works near Sodom, Israel.

Where conventional shaft mining processes are used, the potassium chloride (KCl) is brought to the surface as sylvinite ore. Sylvinite ore is a physical mixture of interlocked crystals of sylvinite (KCl) and halite (NaCl) containing small quantities of dispersed clay and other impurities. The KCl is separated from the NaCl and other minerals by a selective flotation process.

SECONDARY NUTRIENTS

In the fertilizer industry and in agriculture generally, the importance of the major elements, nitrogen, phosphorus and potassium, is well established. Secondary and micronutrients are also fully recognized as essential for plant growth. When guarantees are made for secondary and micronutrients, the usual requirement is that they be stated in terms of the elements. Table 5-5 gives, along with other data, the sulfur and calcium contents of most of the commonly used fertilizers.

Calcium sources for crop nutrition are rarely needed. Foliar sprays for celery, tomatoes and apples have included calcium nitrate. Table 5-5 gives the calcium content of N, P and K sources. Other sources which provide calcium include soil amendments, manure and irrigation water.

Magnesium sources for crop nutrition include Epsom salts (magnesium sulfate, the double salt of potassium-magnesium sulfate and magnesium nitrate used for foliar applications on citrus. On acid soils, to which lime is applied, the use of dolomitic lime provides not only calcium but also magnesium.

Sulfur sources for nutrition of crops are provided in many N, P and K materials. (Consult Table 5-5 for data.) Additional sulfur sources include:

Source	Percent Sulfur
Elemental sulfur	99
Gypsum	16-18
Sulfuric acid (95-99%)	32
Ferrous sulfate	11.5
Ferric sulfate	18-19
Calcium polysulfide solution	25
Ammonium polysulfide solution	40-45
Ammonium bisulfite solution (8.5% N)	17
Ammonium thiosulfate solution (12% N)	26

Other common sources of sulfur include manure, most river water, rain water and pesticidal sulfur.

MICRONUTRIENTS

Properties of the various micronutrient sources vary considerably. A micronutrient material may be completely water soluble or very slightly soluble. Inorganic sources may be relatively pure compounds or mixtures of compounds containing one or more micronutrients, with or without nonmicronutrient compounds. Organic sources are available as synthetic chelates or natural organic complexes of metal ions. Therefore, one classification of micronutrient sources includes: (1) inorganic salts, (2) synthetic chelates and (3) natural organic complexes.

Inorganic Salts

Sulfates of Cu, Fe, Mg and Zn and borates and molybdates are the most common sources of inorganic micronutrients. The most commonly used boron source is sodium tetraborate. This compound is relatively water soluble. Soluble boric acid or sodium octaborate (polybor) is often used as foliar sprays.

Water-soluble ammonium and sodium molybdate are the primary sources of molybdenum. Some molybdic oxide, a slightly soluble compound, is occasionally used.

See Table 5-6 for consumption and properties of inorganic micronutrient materials.

Synthetic Chelates

A chelating agent is a compound (usually organic) which can combine with a metal ion and form a ring structure between a portion of the chelating agent molecule and the metal. This delays the metal ion from precipitating in the soil to form insoluble compounds. Commercially available synthetic chelating agents and the concentration of micronutrients are shown in Table 5-7. The general stability sequence of these chelates of micronutrients and calcium decreases as follows: $Fe^{+++}> Cu^{++}> Zn^{++}> Fe^{++}> Mn^{++}> Ca^{++}$.

Other factors may determine the relative quantities present in a given set of conditions. The relative stability of ferric chelates decreases in the order of EDDHA$>$ DTPA$>$ EDTA$>$ HEEDTA$>$

Table 5-5. Average Composition of Fertilizer Materials

Fertilizer Materials	Chemical Formula	Total Nitrogen N%	Available Phosphoric Acid P₂O₅%	Water-Soluble Potash K₂O%	Combined Calcium Ca%	Combined Sulfur S%	Equivalent Acidity or Basicity in Lbs. CaCO₃*	
							Acid	Base
Nitrogen materials								
Ammonium nitrate	NH_4NO_3	33.5-34					62	
Ammonium nitrate-sulfate	$NH_4NO_3 \cdot (NH_4)_2SO_4$	30				6.5	68	
Monoammonium phosphate	$NH_4H_2PO_4$	11	48				58	
Ammonium phosphate-sulfate	$NH_4H_2PO_4 \cdot (NH_4)_2SO_4$	13	39			7	69	
Ammonium phosphate-sulfate	$NH_4H_2PO_4 \cdot (NH_4)_2SO_4$	16	20			15	88	
Ammonium phosphate-nitrate	$NH_4H_2PO_4 \cdot NH_4NO_3$	27	12			4.5	75	
Diammonium phosphate	$(NH_4)_2HPO_4$	16-18	46-48				70	
Ammonium sulfate	$(NH_4)_2SO_4$	21				24	110	
Anhydrous ammonia	NH_3	82					148	
Aqua ammonia	NH_4OH	20					36	
Calcium ammonium nitrate solution	$Ca(NO_3)_2 \cdot NH_4NO_3$	17			8.8		9	
Calcium nitrate	$Ca(NO_3)_2$	15.5			21			20
Calcium cyanamide	$CaCN_2$	20-22			37			63
Sodium nitrate	$NaNO_3$	16						29
Urea	$CO(NH_2)_2$	45-46					71	
Urea formaldehyde†		38					60	

(Continued)

Table 5-5 (Continued)

Fertilizer Materials	Chemical Formula	Total Nitrogen N%	Available Phosphoric Acid P₂O₅%	Water-Soluble Potash K₂O%	Combined Calcium Ca%	Combined Sulfur S%	Equivalent Acidity or Basicity in Lbs. CaCO₃* Acid	Base
Urea ammonium nitrate solution	$NH_4NO_3 \cdot Co(NH_2)_2$	32					57	
Phosphate materials								
Single superphosphate	$Ca(H_2PO_4)_2$		18-20		18-21	12	neutral	
Triple superphosphate	$Ca(H_2PO_4)_2$		45-46		12-14	1	neutral	
Phosphoric acid	H_3PO_4		52-54				110	
Superphosphoric acid	‡		76-83				160	
Potash materials								
Potassium chloride	KCl			60-62			neutral	
Potassium nitrate	KNO_3	13		44-46				23
Potassium sulfate	K_2SO_4			50-53		18	neutral	
Sulfate of potash-magnesia	$K_2SO_4 \cdot 2MgSO_4$			22	0.1	22	neutral	

* Equivalent per 100 lbs. of each material.
† Also known as ureaform, reaction product of urea and formaldehyde.
‡ H_3PO_4, $H_4P_2O_7$, $H_5P_3O_{10}$, $H_6P_4O_{13}$ and other higher phosphate forms.

Table 5-6.   Inorganic Sources of Micronutrients

Material	Element	Water Solubility	°F
	(%)	(gm/100 gm H_2O)	
Sources of boron			
Granular borax–$Na_2B_4O_7$·$10H_2O$	11.3	2.5	33
Sodium tetraborate, anhydrous–$Na_2B_4O_7$	21.5	1.3	32
Polybor	20.5	22	86
Ammonium pentaborate–$NH_4B_5O_8$·$4H_2O$	19.9	7	64
Sources of copper			
Copper sulfate–$CuSO_4$·$5H_2O$	25.0	24	32
Cuprous oxide–Cu_2O	88.8	i*	
Cupric oxide–CuO	79.8	i*	
Cuprous chloride–Cu_2Cl_2	64.2	1.5	77
Cupric chloride–$CuCl_2$	47.2	71	32
Sources of iron			
Ferrous sulfate–$FeSO_4$·$7H_2O$	20.1	33	32
Ferric sulfate–$Fe_2(SO_4)_3$·$9H_2O$	19.9	440	68
Iron oxalate–$Fe_2(C_2O_4)_3$	30.0	very soluble	
Ferrous ammonium sulfate			
$Fe(NH_4)_2(SO_4)_2$·$6H_2O$	14.2	18	32
Ferric chloride–$FeCl_3$	34.4	74	32
Sources of manganese			
Manganous sulfate–$MnSO_4$·$4H_2O$	24.6	105	32
Manganous carbonate–$MnCO_3$	47.8	0.0065	77
Manganese oxide–Mn_3O_4	72.0	i*	
Manganous chloride–$MnCl_2$	43.7	63	32
Manganous oxide–MnO	77.4		
Sources of molybdenum			
Sodium molybdate–Na_2MoO_4·H_2O	39.7	56	32
Ammonium molybdate–$(NH_4)_6Mo_7O_{24}$·$4H_2O$	54.3	44	77
Molybdic oxide–MoO_3	66.0	0.11	64
Sources of zinc			
Zinc sulfate–$ZnSO_4$·H_2O	36.4	89	212
Zinc oxide–ZnO	80.3	i*	
Zinc carbonate–$ZnCO_3$	52.1	0.001	60
Zinc chloride–$ZnCl_2$	48.0	432	77
Zinc oxysulfate–ZnO·$ZnSO_4$	53.8	—	—
Zinc ammonium sulfate–			
$ZnSO_4$·$(NH_4)_2SO_4$·$6H_2O$	16.3	9.6	32
Zinc nitrate–$Zn(NO_3)_2$·$6H_2O$	22.0	324	68

*i denotes insolubility.

Table 5-7.   Synthetic Chelates

Chelating Agent	Micronutrient Content, Percent Element			
	Cu	Fe	Mn	Zn
EDTA	7-13	5-14	5-12	6-14
HEEDTA	4-9	5-9	5-9	9
NTA	—	8	—	13
DTPA	—	10	—	—
EDDHA	—	6	—	—

NTA, whereas the stability of these chelates of the other ions listed above decreases as follows: DTPA> EDTA> HEEDTA> NTA> EDDHA. In other words, EDDHA chelation of iron is greatly favored over the other micronutrients.

Generally, the stability of metal chelates is greater near neutral than at low or high pH values. This is an important consideration when incorporating metal chelates into macronutrient fertilizers. Mixing ZnEDTA with phosphoric acid prior to ammoniation results in a breakdown of the chelate, but adding ZnEDTA with the ammoniating solution leaves a stable chelate.

Natural Organic Complexes

Many naturally occurring compounds contain chemically reactive groups similar to synthetic chelating agents. Those used commercially in complex micronutrients are usually prepared from by-products of the wood pulp industry. Metal chelates of these compounds have relatively lower stability than the common synthetic chelates. Also, these chelates are more readily broken down by microorganisms in soil. Most are suitable for foliar sprays and mixing with fluid fertilizers.

ORGANIC PRODUCTS

Organic products and organic materials can be classified in several different ways. Strictly speaking, the term *organic* denotes carbon, including that of synthetic origin. However, organic fertilizers are usually considered naturally occurring compounds. In this category fall the wastes from sewage plants and manures. Generally they are good soil amendments. They add quantities of organic matter to the soil along with small amounts of plant nutrients. The user should be aware that sewage and industrial wastes may be contaminated with high levels of toxic elements such as cadmium and lead. Continuous use could cause excessive levels of these toxic elements to enter the crops. Table 8-3 shows the analysis of some of the most common organic products and manures.

Composts of household residues can be disposed of beneficially in improvement of the homeowner's garden. The nutrients contained in the compost can be utilized, and the organic matter will help improve soil structure. Organic materials should not be wasted, but treated as resources. It is, however, economically unsound to expect the commercial farmer to meet his crops' nutrient needs through organic products.

SPECIALTY FERTILIZERS

Considerable effort has been directed toward developing fertilizers to overcome a specific problem, or directed at the nutrient needs of a specific crop. These products are considered specialty fertilizers.

Two main approaches to achieving controlled feeding have been the development of compounds of limited water solubility and altering soluble materials to retard their release to the soil solution. Nitrogen, the most widely used fertilizer element, and the most susceptible to loss by either leaching or denitrification, has received the greatest attention.

Controlling the release of nitrogen can be accomplished by: (1) adding a physical barrier (coating) to water soluble materials; (2) using materials of limited water solubility, e.g., metal ammonium phosphates; and (3) using materials of limited water solubility which during chemical and/or microbiological decomposition release nutrients in available forms, e.g., ureaforms.

Coated Fertilizers

Urea, because of its high analysis and physical properties, is the principal material used in coated fertilizers. One commercial coating process for fertilizers uses a resin (Osmocote®) coating. Sulfur is also used as a coating to slow the release of nitrogen.

Uncoated Inorganic Materials

These compounds comprise a group of compounds of the general formula $MeNH_4PO_4 \cdot xH_2O$ where Me is a divalent metal. All have limited solubility, and several have been developed as multiple nutrient, slow release fertilizers. Magnesium ammonium phosphate is the most common. Water-soluble nitrogen in these materials varies between 1 and 2 percent.

Uncoated Organic Compounds

This category represents the largest quantity of specialty fertilizers on the market. Most use urea since urea reacts with a number of aldehydes to form compounds which are sparingly soluble in water. Two of the most common are the ureaforms (UF) and isobutylidene-diurea (IBDU). Urea reacts with formaldehyde in the presence of a catalyst to form a mixture of compounds under the

Table 5-8. Analyses of Some Specialty Fertilizer Compounds

Material	Method of Controlling Release	Nutrient Content
		(%)
Osmocote®	Coating with resin	18-5-11
		14-14-14
		18-6-12
Sulfur-coated urea	Coating with sulfur	32-37 N
Isobutylidene diurea	Solubility	31 N
Urea formaldehyde	Solubility	35-38 N

generic name *ureaforms*. The composition of ureaforms is controlled by the mole ratio of urea to formaldehyde. In manufacturing, there is a problem of obtaining a suitable balance between the relatively soluble short-chain and more insoluble long-chain polymers. IBDU is the major reaction product of urea and isobutylaldehyde. It is a white compound with low water solubility.

NUTRIENT CONVERSION FACTORS

With the exceptions of phosphorus and potassium, the nutrients in fertilizers are reported in elemental form. Figure 13-1 can be used for rapid conversion of phosphorus and potassium to the oxide forms or the reverse.

Several foreign countries report the nutrients in fertilizers on the elemental basis. Most of the scientific literature is also reported this way. Oftentimes it is necessary to determine the percentage of an element in a fertilizer material. Table 5-9 provides easy conversion factors to make the conversion from the compound to the element and from the element to the compound for common fertilizer materials.

Table 5-9.   Conversion Factors

To find the equivalent of one material A in terms of another B, multiply the amount of A by the factor in column A to B. To find the equivalent of material B in terms of A, multiply the amount of B by the factor in column B to A.

A	B	Multiply A to B	B to A
Ammonia (NH$_3$)	Nitrogen (N)	0.8224	1.2159
Nitrate (NO$_3$)	Nitrogen (N)	0.2259	4.4266
Protein (crude)	Nitrogen (N)	0.1600	6.2500
Ammonium nitrate (NH$_4$NO$_3$)	Nitrogen (N)	0.3500	2.8572
Ammonium sulfate [(NH$_4$)$_2$SO$_4$]	Nitrogen (N)	0.2120	4.7168
Calcium nitrate [Ca(NO$_3$)$_2$]	Nitrogen (N)	0.1707	5.8572
Potassium nitrate (KNO$_3$)	Nitrogen (N)	0.1386	7.2176
Sodium nitrate (NaNO$_3$)	Nitrogen (N)	0.1648	6.0679
Monoammonium phosphate (NH$_4$H$_2$PO$_4$)	Nitrogen (N)	0.1218	8.2118
Diammonium phosphate [(NH$_4$)$_2$HPO$_4$]	Nitrogen (N)	0.2121	4.7138
Urea [(NH$_2$)$_2$CO]	Nitrogen (N)	0.4665	2.1437
Phosphoric acid (P$_2$O$_5$)*	Phosphorus (P)	0.4364	2.2914
Phosphate (PO$_4$)	Phosphorus (P)	0.3261	3.0662
Monoammonium phosphate (NH$_4$H$_2$PO$_4$)	Phosphoric acid (P$_2$O$_5$)*	0.6170	1.6207
Diammonium phosphate [(NH$_4$)$_2$HPO$_4$]	Phosphoric acid (P$_2$O$_5$)*	0.5374	1.8607
Monocalcium phosphate [Ca(H$_2$PO$_4$)$_2$]	Phosphoric acid (P$_2$O$_5$)*	0.6068	1.6479
Dicalcium phosphate (CaHPO$_4$·2H$_2$O)	Phosphoric acid (P$_2$O$_5$)*	0.4124	2.4247
Tricalcium phosphate [Ca$_3$(PO$_4$)$_2$]	Phosphoric acid (P$_2$O$_5$)*	0.4581	2.1829
Potash (K$_2$O)	Potassium (K)	0.8301	1.2046
Muriate of potash (KCl)	Potash (K$_2$O)	0.6317	1.5828
Sulfate of potash (K$_2$SO$_4$)	Potash (K$_2$O)	0.5405	1.8499

(Continued)

Table 5-9 (Continued)

A	B	Multiply A to B	Multiply B to A
Potassium nitrate (KNO$_3$)	Potash (K$_2$O)	0.4658	2.1466
Potassium carbonate (K$_2$CO$_3$)	Potash (K$_2$O)	0.6816	1.4672
Gypsum (CaSO$_4$·2H$_2$O)	Calcium sulfate (CaSO$_4$)	0.7907	1.2647
Gypsum (CaSO$_4$·2H$_2$O)	Calcium (Ca)	0.2326	4.3000
Gypsum (CaSO$_4$·2H$_2$O)	Calcium oxide (CaO)	0.3257	3.0702
Calcium oxide (CaO)	Calcium (Ca)	0.7147	1.3992
Calcium carbonate (CaCO$_3$)	Calcium (Ca)	0.4004	2.4973
Calcium carbonate (CaCO$_3$)	Calcium oxide (CaO)	0.5604	1.7848
Calcium carbonate (CaCO$_3$)	Calcium hydroxide [Ca(OH)$_2$]	0.7403	1.3508
Calcium hydroxide [Ca(OH)$_2$]	Calcium (Ca)	0.5409	1.8487
Magnesium oxide (MgO)	Magnesium (Mg)	0.6032	1.6579
Magnesium sulfate (MgSO$_4$)	Magnesium (Mg)	0.2020	4.9501
Epsom salts (MgSO$_4$·7H$_2$O)	Magnesium (Mg)	0.0987	10.1356
Sulfate (SO$_4$)	Sulfur (S)	0.3333	3.0000
Ammonium sulfate [(NH$_4$)$_2$SO$_4$]	Sulfur (S)	0.2426	4.1211
Gypsum (CaSO$_4$·2H$_2$O)	Sulfur (S)	0.1860	5.3750
Magnesium sulfate (MgSO$_4$)	Sulfur (S)	0.3190	3.1350
Potassium sulfate (K$_2$SO$_4$)	Sulfur (S)	0.1837	5.4438
Sulfuric acid (H$_2$SO$_4$)	Sulfur (S)	0.3269	3.0587
Borax (Na$_2$B$_4$O$_7$·10H$_2$O)	Boron (B)	0.1134	8.8129
Boron trioxide (B$_2$O$_3$)	Boron (B)	0.3107	3.2181
Sodium tetraborate pentahydrate (Na$_2$B$_4$O$_7$·5H$_2$O)	Boron (B)	0.1485	6.7315
Sodium tetraborate anhydrous (Na$_2$B$_4$O$_7$)	Boron (B)	0.2150	4.6502
Cobalt nitrate [Co(NO$_3$)$_2$·6H$_2$O]	Cobalt (Co)	0.2025	4.9383
Cobalt sulfate (CoSO$_4$·7H$_2$O)	Cobalt (Co)	0.2097	4.7690
Cobalt sulfate (CoSO$_4$)	Cobalt (Co)	0.3802	2.6299
Copper sulfate (CuSO$_4$)	Copper (Cu)	0.3981	2.5119
Copper sulfate (CuSO$_4$·5H$_2$O)	Copper (Cu)	0.2545	3.9293
Ferric sulfate [Fe$_2$(SO$_4$)$_3$]	Iron (Fe)	0.2793	3.5804
Ferrous sulfate (FeSO$_4$)	Iron (Fe)	0.3676	2.7203
Ferrous sulfate (FeSO$_4$·7H$_2$O)	Iron (Fe)	0.2009	4.9776
Manganese sulfate (MnSO$_4$)	Manganese (Mn)	0.3638	2.7486
Manganese sulfate (MnSO$_4$·4H$_2$O)	Manganese (Mn)	0.2463	4.0602
Sodium molybdate (Na$_2$MoO$_4$·2H$_2$O)	Molybdenum (Mo)	0.3965	2.5218
Sodium nitrate (NaNO$_3$)	Sodium (Na)	0.2705	3.6970
Sodium chloride (NaCl)	Sodium (Na)	0.3934	2.5417
Zinc oxide (ZnO)	Zinc (Zn)	0.8034	1.2447
Zinc sulfate (ZnSO$_4$)	Zinc (Zn)	0.4050	2.4693
Zinc sulfate (ZnSO$_4$·1H$_2$O)	Zinc (Zn)	0.3643	2.7449

*Also called phosphoric acid anhydride, phosphorus pentoxide, available phosphoric acid.

SUPPLEMENTARY READING

1. *Agricultural Anhydrous Ammonia: Technology and Use.* M. H. McVickar, W. P. Martin, I. E. Miles and H. H. Tucker. Soil Sci. Soc. of America. 1966.
2. *Fertilizer Nitrogen, Its Chemistry and Technology,* Second Edition. V. Sauchelli. Reinhold Publishing Corp. 1968.
3. *Fertilizer Technology and Use,* Second Edition. R. A. Olson, T. J. Army, J. J. Hanway and V. J. Kilmer. Soil Sci. Soc. of America. 1971.
4. The Merck Index, 9th Edition. Merck and Co., Inc. 1978.
5. *Soil Fertility and Fertilizers,* Third Edition. S. L. Tisdale and W. L. Nelson. The Macmillan Company. 1975.
6. *Using Commercial Fertilizers,* Fourth Edition. M. H. McVickar and W. W. Walker. The Interstate Printers & Publishers, Inc. 1978.

Chapter 6

FERTILIZER FORMULATION, STORAGE
AND HANDLING

With the rapid technological developments of recent years in the fertilizer trade, the farmer, as well as the fertilizer dealer and distributor, has a wide choice of fertilizer systems available for use. These may be grouped into three broad classifications. First, is the system that makes direct application of homogeneous products of uniform size and composition. Such products are available, having a wide range of grades and ratios. Storage and handling are usually in bulk form, but bagged products may also be obtained. Second, is a system using products from the first category, along with other fertilizer materials in granular form, for use in bulk blending plants to produce ratios and grades readily adaptable to individual areas and farms. Such products have been generally sized to minimize segregation in storage and handling. The third grouping is the fluid fertilizer system that has products ranging from clear liquid solutions to suspensions. It has as its assets ease of handling, uniform composition and adaptability to additions of herbicides and insecticides.

The adoption of these systems has been hastened because of the development of new products and because they provide cost savings at all levels from manufacture, transportation, storage and application. With new technology breakthroughs, substitution or shift from one system to another will occur. This is well illustrated by the development of superphosphoric acid which has allowed for the expansion of the fluid fertilizer business. Suspension fertilizers can now be formulated that have twice the nutrient content of clear liquids and that give most of the advantages claimed for liquid fertilizers.

The fertilizer supplier has several options as to the type of production unit he operates and the materials he has available. Factors influencing his choice include capital assets, size of distribution area, availability of raw materials, agronomic suitability of certain products and personal preferences.

103

FORMULATION

Homogeneous Products

The unique characteristic of a homogeneous product is that each granule or pellet has the same analysis. Homogeneous fertilizer products are usually manufactured in large factories and are supplied to the resellers where they may be used in blends or applied directly to the land. Homogeneous plants utilize ammonia, sulfuric and phosphoric acids and other raw materials.

Some common grades are 18-46-0, 16-20-0, 16-16-16, 19-9-0, 11-55-0, 27-12-0, 6-20-20 and 12-12-12. In addition to three primary nutrients, nitrogen, phosphate and potassium, it is also possible to include micronutrients as a guaranteed constituent in the product.

Bulk Blends

Bulk blends are physical mixes of two or more fertilizer materials. The bulk blend plant receives fertilizer products from a basic pro-

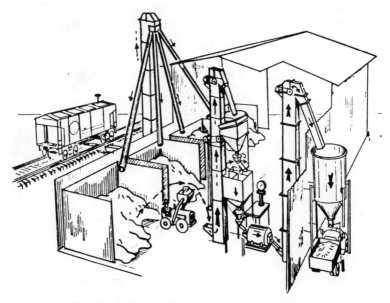

Fig. 6-1.   Schematic drawing of a bulk blending plant.

ducer, stores them and blends them together as needed in some type of mixing device. Some of the materials more commonly used to make blends are ammonium nitrate, ammonium sulfate, diammonium phosphate, urea and potash materials. The blends may be taken directly to the field and spread or, in some cases, they may be bagged. One of the persistent problems with blends is segregation or separation of one component or raw material from another. Frequently this means separation of the nitrogen from the phosphorus or potash. Consequently, when a blend is spread, uneven distribution of nutrients may occur. The principal contributing factor to segregation is the use of materials of uneven size. For example, if fine crystalline ammonium sulfate is blended with granular diammonium phosphate, segregation occurs due to difference in the size of the two materials. Most fertilizers are screened through a $-6+16$ Tyler screen. If one component such as diammonium phosphate is predominantly $-6+8$ and the other, ammonium sulfate, is predominantly $-12+16$, segregation occurs even though both lie within the $-6+16$ range.

Blenders must exercise care in the selection of raw materials. Also, blenders should avoid allowing piles of finished product to cone, either in storage or when loaded into a truck or trailer. In coning, the larger materials run to the outside of a pile; the small materials stay in the center of the pile. Coning can be largely overcome with the use of a flexible spout. Particle shape and density contribute little to segregation problems.

Micronutrients can be added to blends, but a granular form having the same size range as other components is recommended. Another method of incorporating micronutrients is by spraying a solution of the elements on the blend during mixing. The amount that can be applied in this manner is limited by the amount of liquid spray that can be introduced into a dry blend before problems develop.

Liquids and Suspensions

Fluid fertilizers may be classified as liquids and suspensions. Liquids include nitrogen solutions, phosphoric acid and liquid mixes. Nitrogen solutions have already been discussed in Chapter 5.

Liquid mixes are produced by neutralizing phosphoric acid with ammonia. If ortho acid is used, the usual product will be an 8-24-0 grade. Ammoniation can be accomplished in mild steel tanks. Where ammonia and the acid come in contact and are mixed, stainless steel

Table 6-1. Properties of Ammonium Phosphate Liquids

Grade	8-24-0	9-30-0	10-34-0	11-37-0
Acid used in production	Orthophosphoric	Superphosphoric	Superphosphoric	Superphosphoric
Percent N by weight	8	9	10	11
Percent P_2O_5 by weight	24	30	34	37
Density, lbs./gal. @ 60° F	10.5	11.3	11.4	11.7
N content, lbs./gal.	0.84	1.02	1.14	1.29
P_2O_5 content, lbs./gal.	2.52	3.39	3.87	4.33
Polyphosphates, % of total P_2O_5	none	40 - 45	>50	>65
Viscosity, CP @ 75°	—	—	73	80
Safe storage temperature, °F	12	0	below 0	0
pH	6.4 - 6.6	6.2 - 6.6	5.8 - 6.1	5.8 - 6.2

is preferred. Considerable heat is released during ammoniation, and the solution will be hot. If potash is desired, it should be added at this time, as the introduction of potash will lower the solution temperature.

If superphosphoric acid (68 to 76 percent P_2O_5, 50 to 75 percent of which is present as polyphosphate) is ammoniated, it is possible to produce a stable 10-34-0 solution. The higher analysis of this solution is a result of the greater solubility of the pyro, tripoly and other more condensed phosphates. As condensed species are rather unstable at high temperatures, it is necessary to cool 10-34-0 during production. At high temperatures hydrolysis of the condensed phosphate occurs with reversion to the ortho form as illustrated in Figure 6-2. When this occurs, the "salting out" temperature is raised, making the solution unstable. Therefore, cooling the solution to below 100°F within an hour after production is recommended.

Low poly (20 percent) superphosphoric acid contains significant quantities of heavy metal ions such as magnesium, iron and aluminum. When the acid is ammoniated in the tee-pipe reactor process, developed at TVA, the heat of reaction between the acid and the ammonia drives off additional combined water and forms a large proportion of condensed phosphate (70 to 80 percent) in the final 10-34-0 product. A sketch of a pipe reactor unit is shown in Figure 6-3.

Heavy metal salts of long-chain condensed phosphates are more soluble than ortho and pyro salts. Consequently, the heavy metals are in solution and do not cause precipitation or sludge problems in the 10-34-0 grade.

Although rather high analysis N–P grades can be produced as

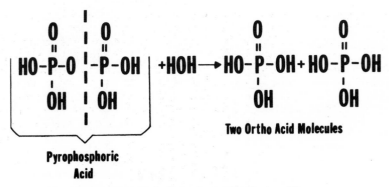

Fig. 6-2. Hydrolysis of pyrophosphoric acid.

clear liquids, the addition of potash raises the "salt out" temperature, thus making it impossible to produce high analysis N–P–K grades. There has been a constant increase in the grade of fluids with the progression from orthophosphate to polyphosphate systems. With the escalation of energy costs, the availability of superphosphoric acid is diminishing, making high analysis, clear liquid fertilizers less available. It thus appears that suspensions will play a more significant role in the fluid fertilizer market in the future.

Suspensions are saturated solutions with crystals of plant nutrients or other materials suspended in the solutions. Usually, a suspending agent such as an attapulgite-type gelling clay is used to suspend the undissolved salt crystals throughout the liquid medium.

Some of the advantages suspensions have over liquid mix fertilizers are:

1. Higher analysis suspension grades can have as much as twice the nutrient content of clear liquid mixtures.
2. Larger amounts of micronutrients can be used in suspensions.

Typical grades of clear liquid mixes and suspension fertilizers are shown in Table 6-2.

Table 6-2. Typical Grades of Some Clear Liquid Mixes and Suspension Fertilizers for Different Ratios

	Grade	
Ratio	Clear Liquid Mix	Suspension
3:1:0	24-8-0	27-9-0
2:1:0	22-11-0	26-13-0
1:1:0	19-19-0	21-21-0
1:1:1	8-8-8	15-15-15
1:2:2	5-10-10	10-20-20
1:3:1	7-21-7	10-30-10
1:3:2	5-15-10	9-27-18
1:3:3	3-9-9	7-21-21

Suspensions can be produced through four principal routes: combining orthophosphoric acid with ammonia, combining polyphosphoric acid with ammonia, combining solid ammonium phosphates with ammonia, and combining diammonium phosphate with sulfuric acid.

A recent development in suspension production has been the use of dry ammonium phosphates, particularly monoammonium phosphate.

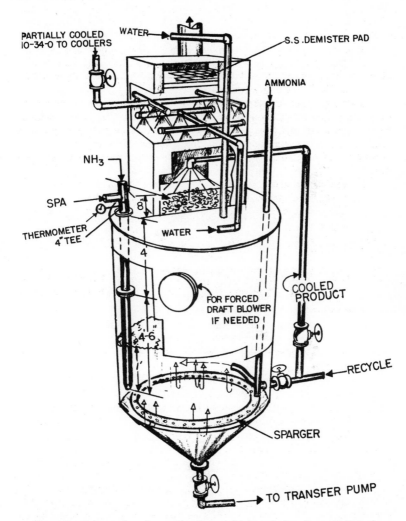

Fig. 6-3. Reactor for the production of high polyphosphate 10-34-0 from low polyphosphoric acid.

This system may have certain advantages, as it permits the producer to use inexpensive dry material storage, for the phosphate in the ammonium phosphate may be less costly than the phosphate in phosphoric acid. The dry ammonium phosphate also permits a dealer to market both dry and liquid materials with only a modest increase in capital cost.

To produce suspensions from monoammonium phosphate, one must have a mixing tank, an agitator and a means of sparging in ammonia and introducing clay. An individual must understand the relationship between ammonium nitrogen and P_2O_5 to produce good quality suspensions.

There may be application problems with suspensions. Due to the presence of the finely divided salt crystals in the fluid, buildup and clogging can occur. Constant agitation is needed to keep the materials in suspension.

STORAGE AND HANDLING

To bring efficiency to fertilizer production and distribution, proper and safe storage must be an integral part of the system. Fertilizer plants should operate throughout the year to be efficient. Storage by the producer is necessary to keep his facilities operating, and storage by the dealer and farmer is necessary to insure having a supply available when needed. Fertilizer storage will not be elaborated on here, but those who contemplate storage should make a thorough study of their needs and the facilities available.

Dry Materials

Ammonium nitrate is an excellent fertilizer material that presents no hazard where good storage and proper handling procedures are observed. Precautions to be taken are:

1. Keep it away from open flame.
2. Avoid contaminating it with foreign matter.
3. In case of fire, flood the area with water.
4. Burn empty bags out of doors.
5. Sweep up and dispose of all contaminated material.
6. Do not store in close proximity to steam pipes or radiators.
7. Keep it separate from other materials stored in the same warehouse, especially combustible materials and urea.

Dry urea is totally incompatible with dry ammonium nitrate.

Some ammonium phosphates contain small amounts of ammonium nitrate, and the mixing of these materials with urea should be avoided. Ammonium nitrate has a critical relative humidity of 59.4 at 86°F, and under humid conditions it tends to absorb moisture. In areas of high humidity where ammonium nitrate is manufactured, dehumidified storage may be necessary. At small retail outlets and bulk blend plants, a tight bin and a polyethylene cover sheet can be used for storage.

Although ammonium nitrate has been used safely for many years, one should be aware that it can be converted into an explosive agent. To make it explosive, it is necessary to mix it with some type of organic material, the most common being diesel fuel. Whenever ammonium nitrate accidentally becomes contaminated with fuel oil or other combustible materials, it should be disposed of in a safe manner.

Urea—The physical handling and storage of urea is very similar to ammonium nitrate. Urea is considered nonexplosive and does not have the restrictive regulations imposed on ammonium nitrate. Its critical relative humidity is 72 at 86°F. Thus, it too is hygroscopic.

Ammonium sulfate—This material is safe and easy to store. Because of its high critical relative humidity of 81 at 86°F, storage problems are infrequent. The product is corrosive, so concrete or wooden storage structures are preferred.

Phosphates and potash—Except under extremely adverse conditions the ammonium phosphates, straight phosphates and potash materials require no specialized storage. Like most fertilizers, they tend to be corrosive, so concrete and wood are preferred materials for storage structures. Due to the density, large piles or bags of these materials stacked excessively high may cause "setting up," but the lumps are easily broken.

Liquid Materials

Anhydrous ammonia—This material is widely used in manufacturing and for direct application. It is potentially hazardous; therefore, proper safety precautions must be observed. Handling procedures and safety precautions are well known. Anhydrous ammonia storage and handling will not be discussed here, but it is recommended that anyone contemplating handling or using this product should acquaint himself with its characteristics and proper handling procedures. A good source of information is The Fertilizer Institute's manual, *Agricultural Ammonia Safety*.

Aqua ammonia—A 20 percent solution of aqua does not have a gauge pressure, but ammonia vapor is constantly leaving the solution. Therefore, a pressure-vacuum valve must be installed on the tank. This is also true with other solutions which contain free ammonia.

Urea–ammonium nitrate solutions—Two commonly used non-pressure solutions made from urea, ammonium nitrate and water are standardized at 32 percent nitrogen and 28 percent nitrogen. The latter is used during cold weather since it has a lower salting out temperature. These solutions are used for direct application and for making multinutrient liquid fertilizers by combining them with neutral phosphate solutions and potash, if desired.

Ammonium nitrate solution—The common non-pressure solution of ammonium nitrate in water is usually standardized at 20 percent nitrogen content. It is used for direct application or for making multinutrient liquid fertilizers. Some solutions containing higher concentrations of ammonium nitrate are used for manufacturing purposes only since they need to be kept hot to prevent salting out.

Phosphoric and superphosphoric acids—These acids are widely used by both the fertilizer manufacturer and the farmer. Both are corrosive, although superacid is somewhat less corrosive than orthophosphoric acid. Rubber-lined or 316 stainless steel is necessary for ortho acid and is recommended for superacid.

Superphosphoric acid is hygroscopic, and, if allowed to absorb moisture, will form a thin layer of corrosive ortho acid on the surface. This should be prevented by having a silica-gel breather installed to prevent moisture from entering the tank.

Because of the viscosity characteristics of superacids, they should be handled at temperatures above 130°F. The relationship between phosphoric acid concentration, temperature and viscosity is shown in Figure 6-4.

Although centrifugal pumps can move hot superacid, most operators prefer positive pumps of the gear, screw or positive vane types. Wetted parts should be made of corrosion-resistant alloy. Mechanical seals can be used if the acid is clean. Long lines should be provided with external jacketing or steam tracing, to prevent blockages due to gelling or freezing.

Precautions must be taken against phosphoric or superacid coming in contact with the skin or eyes. These acids are not extremely hazardous, and prompt washing with copious quantities of water is an effective remedy. Superacids of 76 to 83 percent P_2O_5 are strong

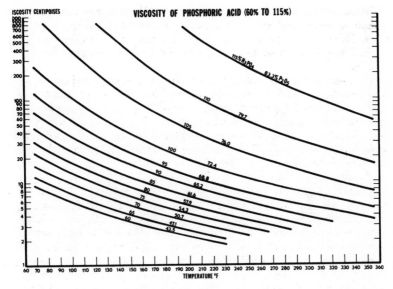

Fig. 6-4. Viscosity of phosphoric acid as affected by temperature and concentration.

dehydrating agents; therefore, they have a greater tendency to cause blistering than the less concentrated acids.

Clear liquid and fluid suspensions—Clear liquid fertilizers are easy to handle. If there are neutral solutions, mild steel storage can be utilized and any conventional pumping system can be used.

Suspensions store best in vertical mild steel or plastic tanks equipped for air sparging. The life of a steel tank can be extended by coating the inside with a material such as an epoxy base paint. The tanks may have flat or cone bottoms, although it is easier to resuspend materials that settle in cone-bottom tanks. When suspensions are not of good quality, solids may tend to collect on the bottom outside walls of a flat-bottom tank where they cannot be resuspended by air sparging.

Air sparging is accomplished when an air sparging ring is mounted on the bottom of the storage tank in such a manner that agitation will occur throughout the tank when the sparging lines are pressurized. The holes in the sparging ring point downward. A compressor with a 300-gallon tank pressurized to 100 to 125 pounds/

Fig. 6-5.   A modern fertilizer manufacturing unit producing dry and liquid fertilizers.

in^2 should be adequate for most storage. Sparging once or twice a week should be adequate for most suspensions. Pumps for suspensions should be of a positive displacement type.

Sulfuric acid—This acid is used extensively in the fertilizer industry. It is mostly used by basic producers in the manufacture of phosphatic compounds and ammonium sulfate. Occasionally, it may be used by a dealer or farmer to deal with a specific soil problem.

Sulfuric acid has inherent hazards, yet it can be handled and used with complete safety by following simple precautionary methods. Use of safe wearing apparel, protective equipment and properly engineered handling and storage equipment makes it possible to operate without accident.

Storage areas should have facilities for drainage and for washing off spills. Where convenient, the use of crushed limestone as a foundation under storage tank areas is considered a good practice because it provides good natural drainage and neutralizes any spills. Storage tanks should be equipped with vents to maintain the tanks at atmospheric pressure. Metal catwalks should be provided for working on top of tanks.

Although sulfuric acid is not flammable, it should not be stored near organic materials, nitrates, carbides, chlorates or metal powders. Contact between concentrated sulfuric acid and these materials may cause ignition. As the strength of acid decreases, the product becomes more corrosive, and the use of a nonreactive liner for the storage tank is advisable.

Sulfur materials for formulation of liquids—These materials include ammonium bisulfite and ammonium thiosulfate. For more details on these materials, see Chapter 5.

A major problem of liquid fertilizers is corrosion, and this demands familiarity with the characteristics of various solutions and the metallic composition of storage containers. Storage vessels, pipelines, valves and fittings of all kinds should not contain materials that are destroyed rapidly when they come in contact with the solution.

SUPPLEMENTARY READING

1. *Agricultural Ammonia Safety.* The Fertilizer Institute. 1974.
2. *Agricultural Anhydrous Ammonia: Technology and Use.* M. H. McVickar, W. P. Martin, I. E. Miles and H. H. Tucker. Soil Sci. Soc. of America. 1966.
3. *Fertilizer Nitrogen, Its Chemistry and Technology,* Second Edition. V. Sauchelli. Reinhold Publishing Corp. 1968.
4. *Fertilizer Technology and Use,* Second Edition. R. A. Olson, T. J. Army, J. J. Hanway and V. J. Kilmer. Soil Sci. Soc. of America. 1971.
5. *Soil Fertility and Fertilizers,* Third Edition. S. L. Tisdale and W. L. Nelson. The Macmillan Company. 1975.
6. *Storage of Ammonium Nitrate.* NFPA No. 490. National Fire Protection Association. 1974.
7. *Using Commercial Fertilizers,* Fourth Edition. M. H. McVickar and W. W. Walker. The Interstate Printers & Publishers, Inc. 1978.

Chapter 7

CORRECTING PROBLEM SOILS
WITH AMENDMENTS

The soils under consideration in this chapter are problem soils that require special management and remedial measures. The primary concern will be with the use of amendments such as lime, gypsum, sulfur and other amendment products which, when properly used, make the soil more productive.

Those soils which have excess acidity may produce toxicity in growing crops from solubilized aluminum and manganese, alter the populations and activities of the microorganisms involved in nitrogen, sulfur and phosphorus transformations in the soil and thus affect the availability of nutrients to higher plants. Acidity may affect the growth of plants directly. Usually it is an indication of a low level of calcium and magnesium. The use of lime as an amendment can ameliorate many or all of these problems.

Those soils which have a high pH, often referred to as alkaline soils, may be afflicted with high sodium content, excess salts, poor structure and other problems. The treatment of these soils with amendments requires a completely different approach. This chapter will be concerned with the identification of the problems and the appropriate measures to take in their correction.

ACID SOILS

In the West, acid soils are generally found in areas of heavy rainfall, on sandy or light textured soils, where high rates of acid-forming fertilizers have been used on poorly buffered soils and on peat or organic soil deposits. Acid soils are more readily formed from acid igneous rocks such as granites and secondary rocks such as sandstones, than from basic igneous rocks such as basalts.

Soils become acid because the cations of the soil colloids, primarily calcium, are replaced by hydrogen ions. This implicates ion exchange and other adsorptive reactions that are associated with the colloids, such as the type of clay, the amount of organic matter, etc. Aluminum also comes into play to the extent that highly acid soils can result from aluminum saturation.

The nature of soil acidity is complex; one part, called the active acidity, is made up of the hydrogen ions in the soil solution. These are the ions measured when the pH of a soil is determined. Another and much larger part of the total acidity is usually referred to as the potential acidity representing the hydrogen ions held on the colloidal surface. Since clays and organic particles have large surface areas, they usually have a much higher total acidity than sandy soils, and it takes more lime to change the pH of these soils.

The pH measurement is commonly made when testing soils. Although it is a useful index, it is not completely understood and is often misused. The measurement is usually made on a 1:1 mixture of soil and water or on a saturated soil paste. Differences of several tenths of a pH unit higher are observed as the water:soil ratio is increased. The interpretation of the pH value must be based upon more than a simple reading.

The pH range for acid soils ordinarily is from 4.0 to 7.0. Values below 4.0 are obtained only when free acids are present, such as sulfuric acid. Values above 7.0 indicate alkalinity.

Lime Requirement

Different methods have been developed to determine the amount of lime needed to bring the pH of an acid soil to a desirable range. All of those presently used take into consideration the soil texture and organic matter content and use a specialized procedure. Table 7-1 shows what the effect of finely ground limestone is on different soils. See also Appendix Table B-21.

Table 7-1. Approximate Amount of Finely Ground Limestone Needed to Raise the pH of a 7-Inch Layer of Soil*

Soil Texture	Lime Requirements (Tons per Acre)	
	From pH 4.5 to 5.5	From pH 5.5 to 6.5
Sand and loamy sand	0.5	0.6
Sandy loam	0.8	1.3
Loam	1.2	1.7
Silt loam	1.5	2.0
Clay loam	1.9	2.3
Muck	3.8	4.3

*Adapted from USDA Agricultural Handbook No. 18.

Liming Materials

The commonly used materials for liming soils are the carbonates, oxides, hydroxides and silicates of calcium and magnesium. A material does not qualify as a liming compound just because it contains calcium and magnesium. Gypsum, for example, has little direct effect on soil pH and cannot be used to correct a low soil pH. Table 7-2 gives the liming materials most commonly used to treat acid soils.

Table 7-2. Common Liming Materials*

Name	Chemical Formula	Equivalent % $CaCO_3$	Source
Shell meal	$CaCO_3$	95	Natural shell deposits
Limestone	$CaCO_3$	100	Pure form, finely ground
Hydrated lime	$Ca(OH)_2$	120-135	Steam burned
Burned lime	CaO	150-175	Kiln burned
Dolomite	$CaCO_3 \cdot MgCO_3$	110	Natural deposit
Sugar beet lime †	$CaCO_3$	80-90	Sugar beet by-product lime
Calcium silicate	$CaSiO_3$	60-80	Slag

*Ground to same degree of fineness.
†Four to 10% organic matter.

The value of a liming material is governed by molecular composition, purity and degree of fineness. Some established standards indicate that all material must pass through a 60-mesh screen to have a full efficiency rating.

The benefits of liming acid soils are broad in scope. Calcium and magnesium are essential plant nutrients, and their addition may provide direct value. Additionally, the correction of chemical, physical and biological conditions may result in striking improvements in plant growth. Phosphorus availability is greatest at a soil pH of 6.5 to 7.0. Toxicities of elements are minimized at pH values of 6.0 to 7.0, and availability of the micronutrients is optimized. Biological activity is favored at a neutral or near neutral pH, including such processes as nitrification, nitrogen fixation, decomposition of plant residues, etc. At the same time, soil aggregation and good structural development are favored.

SALINE AND ALKALI SOILS

Saline soils generally occur in regions with arid or semiarid climates. In humid areas, rainfall is usually sufficient to move the soluble salts out of the soil into drainage waters, making the incidence of salinity rare. However, the intrusion of sea or brackish water may induce soil salinization, especially in low-lying areas. Arid regions are frequently inadequately drained and are subject to high evaporation rates, thus allowing salt buildup to occur. Often irrigation is practiced where there are no drainage outlets, and this results in salinization and alkalization.

The soluble salts that occur come indirectly from the weathering of primary minerals and from waters which carry salts from other locations. For example, the Colorado River at Yuma carries more than 1 ton of salt per acre-foot of water. Using this water will result in rapid buildup of salt unless adequate drainage is provided and proper irrigation practices are used.

Alkali soils contain excessive amounts of sodium. When soils come in contact with waters containing a high proportion of sodium, this cation becomes dominant in the soil solution and replaces calcium and magnesium on the clay. As a consequence of this adsorption of sodium, alkali soils are formed.

Fig. 7-1.   Salt buildup in a California vineyard.

Fig. 7-2. Even a salt tolerant crop such as barley may be affected by salt accumulation.

SALINE SOILS

This term is applied to soils which have a conductivity of the saturation extract greater than 4.0 ms/cm, and the exchangeable sodium percentage is less than 15 percent. These soils normally have a pH value below 8.5 and have good physical properties. Saline-alkali soils sometimes contain gypsum. If this is the case, leaching will help dissolve calcium, and the soil will provide its own gypsum amendment. Free calcium carbonate (limestone) may occur in soil. The selection of amendments in this case can be made from those that dissolve the calcium carbonate and form calcium sulfate right in the soil. Such amendments are soil sulfur, sulfuric acid, ferric and ferrous sulfate, aluminum sulfate, etc.

NONSALINE-ALKALI SOILS (SODIC SOILS)

This term is applied to soils which have conductivity of the saturation extract below 4.0 ms/cm, and the exchangeable sodium percentage is greater than 15 percent. They normally have a pH

Fig. 7-3. Spreading finely ground sulfur as a soil amendment.

value greater than 8.5 and are characterized particularly by their poor physical structure. Alkali (sodic) soils contain sufficient exchangeable sodium to interfere with the growth of most crops. These soils are commonly termed *alkali, black alkali* and *slick spot* soils. The darkened appearance is caused by the dispersed and dissolved organic matter deposited on the soil surface by evaporation.

Alkali (adsorbed sodium) causes disintegration of the soil aggregates, dispersing the soil particles and effectively reducing the large pore space. This makes leaching difficult since the soil becomes almost impervious to water.

On non-calcareous soils treatment must be made with gypsum or other soluble calcium salts. On calcareous soils, treatment may be with gypsum or acidifying materials. Calcium replaces sodium on the clay surface and helps bring about a better physical condition that will allow sodium and excess salts to be leached. Organic materials such as manure, crop residues, etc., may be helpful by providing a better physical condition for leaching.

Reclamation of alkali land must be measured in terms of time and cost of the treatment, the value of crops to be raised and the ultimate value of the land. Because of the special soil conditions

Fig. 7-4.   Using a gypsum pit for treating high sodium water.

which may exist it is advisable to consult with soil specialists for specific recommendations to fit the individual case.

SOIL AMENDMENTS

Identification of the specific problems is necessary before the amendment is chosen. Often efforts are made to amend a saline or alkali soil without awareness that the problem may be compounded by high boron content in the soil or the leaching waters. Also, there are those who claim some special material or process will magically cure the problem. The principles concerning the use and selection of soil amendments are well known, and no shortcut to the proper application of these principles will bring any effective or lasting benefits.

Soil Considerations

The presence of lime (free calcium carbonate) in the soil allows the widest selection of amendments. To test for this, a simple procedure can be followed by taking a spoonful or clod of soil and dropping a few drops of muriatic or sulfuric acid on it. If bubbling or fizzing occurs, this indicates the presence of carbonates or bicarbonates.

If the soil contains lime, any of the amendments listed in Table 7-3 may be used. If lime is absent, select only those amendments containing soluble calcium.

Table 7-3. Commonly Used Materials and Their Equivalent Amendment Values

		Tons of Amendment Equivalent to	
Material (100% Basis)	Chemical Formula	1 Ton of Pure Gypsum	1 Ton of Soil Sulfur
Gypsum	$CaSO_4 \cdot 2H_2O$	1.00	5.38
Soil sulfur	S	0.19	1.00
Sulfuric acid (conc.)	H_2SO_4	0.61	3.20
Ferric sulfate	$Fe_2(SO_4)_3 \cdot 9H_2O$	1.09	5.85
Lime sulfur (22% S)	CaS_x	0.68	3.65
Calcium chloride	$CaCl_2 \cdot H_2O$	0.86	—
Calcium nitrate	$Ca(NO_3)_2 \cdot H_2O$	1.06	—
Aluminum sulfate	$Al_2(SO_4)_3$		6.34

The percent purity is given on the bag or identification tag.

Types of Amendments

Calcium-containing amendments, such as gypsum, react in the soil as follows:

gypsum + sodic soil ⟶ calcium soil + sodium sulfate

Leaching is essential in removing the sodium salt, the amount dependent upon the severity of the alkali problem.

Acids such as sulfuric acid require two steps:

1. sulfuric acid + lime ⟶ gypsum + carbon dioxide + water
2. gypsum + sodic soil ⟶ calcium soil + sodium sulfate

The acid-forming materials such as sulfur go through these steps. First, oxidation:

1. sulfur + oxygen + water ⟶ sulfuric acid
2. sulfuric acid + lime ⟶ gypsum + carbon dioxide + water
3. gypsum + sodic soil ⟶ calcium soil + sodium sulfate

Effectiveness of Amendments

The values given in Table 7-3 are for 100 percent pure amendments. If an amendment is not pure, a simple calculation will indicate the amount needed to be equivalent to 1 ton of pure material:

$$\frac{100}{\% \text{ purity}} = \text{tons}$$

Example: If gypsum is 60 percent pure, the calculation would be $\frac{100}{60} = 1.67$ tons, or 1.67 tons of 60 percent gypsum would be equivalent to 1.00 tons of 100 percent pure gypsum.

When considering sulfur, the purity and degree of fineness must be taken into account. Most sulfur is over 99 percent pure. Sulfur must be oxidized before it is effective as an amendment. The finer the material, the faster it will be oxidized in the soil since greater surface area is exposed.

MANAGEMENT OF SALINE AND ALKALI SOILS

Often it is not practical to completely reclaim saline or alkali soils or even to maintain these soils at a low saline or alkali condition. The reasons may be cost of reclamation, inability to adequately drain, high cost of amendments, low quality irrigation water, etc.

Management practices that aid in the control of salinity and alkalinity include:

1. Selection of crops or crop varieties that have tolerance to salt or alkali.
2. Use of special planting procedures that minimize salt accumulation around the seed.
3. Use of sloping beds or special land preparation procedures and tillage methods that provide a low salt environment for the germinating seed.
4. Use of irrigation water to maintain a high water content to dilute the salts or to leach the salts out of the germination and root growing zone.
5. Use of physical amendments for improving soil structure.
6. Deep ripping the soil to break up hardpan or other impervious layers to provide internal drainage.
7. Use of chemical amendments as described.

It is essential to know the nature of soil, both physical and chemical, the quality and quantity of irrigation water available, the

Fig. 7-5. Wise usage of soil amendments and fertilizers produces high yielding crops.

climate of the area including the growing season, the economics of the situation, etc., before a satisfactory management program can be developed. Consulting with the appropriate authorities and having suitable tests made are essential steps in management. Agricultural Handbook No. 60, a publication of the USDA, is a valuable source of information.

SUPPLEMENTARY READING

1. *Diagnosing Soil Salinity*. USDA Agricultural Information Bulletin 279. 1963.
2. *Diagnosis and Improvement of Saline and Alkali Soils*. USDA Agricultural Handbook No. 60. 1954.
3. *Gypsum and Other Chemical Amendments for Soil Improvement*. University of California, Extension Leaflet 149. 1962.
4. *Reclaiming Saline and Alkali Soils*. Extension Publication, Fresno. 1972.
5. *Soil Acidity and Liming*. Agronomy Monograph 12. Amer. Soc. Agronomy. 1967.
6. *Soil Survey Manual*. USDA Agricultural Handbook No. 18. Soil Survey Staff, U.S. Government Printing Office. 1951.
7. *Soil, Yearbook of Agriculture*. U.S. Government Printing Office. 1957.

Chapter 8

SOIL ORGANIC MATTER

Organic matter is one of the major keys to soil productivity. Western soils differ considerably in organic matter content from region to region. Since western climate is mostly semiarid, the average organic matter content is quite low, usually less than one percent. Soil organic content in the higher rainfall areas, having an abundance of vegetation, may range upwards to 10 or 15 percent. Sacramento Delta soils containing centuries-old accumulations of aquatic vegetation, reeds and sedges may be predominantly organic soils (peat and muck).

Soil organic matter consists of plant and animal residues in various stages of decay (decomposition), living soil organisms and substances synthesized by these organisms. The amount of organic

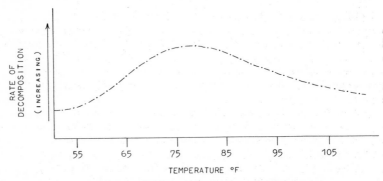

Fig. 8-1.   Temperature influence on the rate of soil microbial activity.

matter that may accumulate in a soil from plant tissues depends upon the temperature, moisture, aeration, soil pH and the amount and chemical nature of the plant residue returned to the soil.

The chemical composition of soil organic matter is categorized in three major groups: *polysaccharides*, *lignins* and *proteins*. Besides

126

these three groups, a variety of other substances such as fats and waxes occur in plant residues. The polysaccharides include cellulose, hemicelluloses, sugars, starches and peptic substances. Lignins are complex materials derived from woody tissues of plants. Protein, which is the principal nitrogen-containing constituent of organic matter, exists in all life forms. These three classes of materials are sources of food for soil microorganisms.

Table 8-1. Common Constituents of Soil Organic Matter and Relative Rates of Decomposition

Organic Constitutents	Approximate Percent of Total Organic Matter	
		Rapidly decomposed
Sugars, starches, simple proteins	1-5	↑
Crude proteins	5-20	
Hemicelluloses	10-25	
Cellulose	30-50	↓
Lignins, fats, waxes	10-30	Very slowly decomposed

Organic residue is decomposed by living organisms in the soil, primarily bacteria, fungi, and actinomycetes. Each group becomes a dominant factor at various stages of decomposition. These and larger organisms such as earthworms and insects ingest organic residue and soil, thereby affecting the binding together of soil particles into stable aggregates. Practically every soil property is affected by soil organic matter.

The dark-colored organic residue which resists further decomposition is referred to as *humus*. Humus increases friability of soils, improves tilth and facilitates aeration and water penetration.

PRINCIPAL BENEFITS OF SOIL ORGANIC MATTER

1. Helps build stable soil aggregates, thus improving soil structure and tilth.
2. Improves aeration and water penetration.
3. Improves moisture-holding capacity.
4. Provides an abundance of negatively charged colloidal-size particles (humus) capable of holding and exchanging nutritive cations.

5. Acts as a buffering agent by decreasing the tendency for an abrupt pH change in the soil when acid or alkaline-forming substances are added.

6. Affects the formation of metal-organic complexes, thus stabilizing soil micronutrients that otherwise might not be available.

7. Provides a source of plant nutrients, particularly nitrogen.

It is important to recognize that the major plant nutrients do not exist in soil organic matter in sufficient quantity to sustain maximum crop growth. Soil organic matter usually contains 5 or 6 percent nitrogen and lesser amounts of phosphorus. These nutrients must be mineralized to the inorganic form during decomposition before they become available. There is some evidence that small quantities of organic compounds can be absorbed directly by plants without mineralization. The contribution is small, however, in the overall nutrition of crops. A simplified equation showing the mineralization process is:

$$\text{organic matter} + H_2O + O_2 \xrightarrow{\text{(soil organisms)}} CO_2 + H_2O + NH_3 + H_3PO_4 + H_2S + \text{other gases} + \text{energy}$$

The carbon in organic matter is the source of energy for soil organisms as they multiply and carry on their life processes. Ammonia produced during mineralization may be nitrified to NO_3 (nitrate) by nitrifying bacteria.

Soil microbes require nutrient elements just as do plants. In the process of breaking down an abundant supply of organic matter, a rapidly growing population of microbes will rob the soil of available nitrogen. This may temporarily reduce plant growth if the soil supply of nitrogen is not sufficient to take care of the needs of both the microbes and growing plants. By adding nitrogen fertilizer, both the growing crop and soil organisms can have a plentiful supply to meet their needs.

THE RECYCLING OF PLANT NUTRIENTS

The recycling of nitrogen from organic matter to soil to growing plants is a part of the soil nitrogen cycle.

The effect of soil organic matter on releasing available phosphorus to growing plants is also significant. The phosphorus cycle is similar in some respects to the nitrogen cycle. For example, when the carbon:phosphorus ratio is wide, immobilization of available phosphorus occurs. When the ratio is narrow, net increases of available inorganic phosphorus are produced.

Humus adsorbs phosphate ions to its surface. These phosphate ions are, however, more available to growing plants than phosphorus precipitated as insoluble compounds. Thus, soils high in humus usually contain greater quantities of available phosphorus than soils low in humus.

SOURCES OF ORGANIC MATERIALS

Organic matter in most agricultural soils is derived from crop residues and manures. Residues which contain the lowest amount of carbon in relation to nitrogen come from cover crops (green manure): legumes, grasses and mustards. These crops decompose rapidly and provide nutrients in excess of microbial needs. Animal manures can also be valuable sources of humus; but, since they contain salts, precautions must be taken when they are used to avoid excessive accumulations of these salts.

Residues of stubble from grain, corn and cotton are low in nitrogen but high in carbon, and thus take longer to decompose. The addition of nitrogen fertilizer to these residues speeds up decomposition and helps to satisfy the microbial demand for the nutrients. When these residues are incorporated, a rule-of-thumb is to apply 20 pounds of nitrogen per ton of residue.

Where sawdust, grain straw and other similar low-nitrogen mulching materials are incorporated, chemical nitrogen should be added to bring the total nitrogen level to $1\frac{3}{4}$ percent by weight of the organic material used. It may not be necessary to add extra nitrogen to composts or organic materials with nitrogen contents above this level.

ORGANIC WASTES

Currently, organic matter in the form of animal manures and sewage sludge is being considered as a source of agricultural nitrogen. This has been brought about through the combination of the necessity for disposing of increasing concentrations of feedlot and other organic waste residues and the shortage of mineral fertilizers. Research has been done to evaluate the effects of applying large quantities of organic wastes, and this means of supplying plant nutrients is often justified and practical.

The yearly rate of nitrogen mineralization (previously discussed)

from any particular organic source is referred to as a decay series.*
This is expressed as the percentage of mineralization occurring each
year following a single application of the organic waste. For instance,
a decay series of 0.30, 0.10, 0.05, means that for any given application, 30 percent is available (mineralized) the first year, 10 percent
of the residual (that which was not previously mineralized) is mineralized the second year, and 5 percent of the residual is mineralized
the third and all subsequent years. Knowing the percent nitrogen of
an organic waste and its decay series should enable growers to
approximate the amount of residue to apply per acre per year to
provide a specific constant amount of nitrogen each year.

Table 8-2 presents data for the total nitrogen inputs required to
maintain a yearly mineralization of 200 pounds of nitrogen per acre
per year for a 20-year period for two decay series for each of six
types of wastes. The first decay series listed in each case is the more
conservative in that the final member of the series is 0.03 or 0.04,
which is considered to be appropriate for colder climates where
decay would be slower.

If organic wastes completely replaced inorganic nitrogen sources,
soluble salts would accumulate when higher concentrations of manure
were applied the first few years. Corrective measures would have to
be taken, and in some moderately saline soils, yields would probably
be reduced.

Until such time that this approach is more completely evaluated
for agricultural production, this information might be useful in
planning waste disposal projects or for the utilization of organic
nitrogen sources in other ways.

ORGANIC CONCENTRATES

Activated sewage sludge, tankage, blood, fish and seed meals are
referred to as organic concentrates and are used more as feeds or
specialty fertilizers than as mulches or organic amendments because
of their high cost. They contain a higher percentage of nitrogen than
most crop residues. Most of the nitrogen is water-insoluble and is
made available by microorganisms during decomposition. Synthetic
organic fertilizers such as urea-formaldehyde condensates contain
significantly higher percentages of nitrogen than natural organic
concentrates. The process of nitrogen release and availability from

* "Using Organic Wastes as Nitrogen Fertilizers." P. F. Pratt, F. E. Broadbent and J. P. Martin. *California Agriculture.* Vol. 27, No. 6, 1973, pp. 10-12.

Table 8-2. Total N Input Required to Maintain a Yearly Mineralization Rate of 200 Pounds per Acre per Year Through a 20-Year Period for Two Decay Series for Each of Six Types of Waste.*

Material and Decay Series	Time, Years							
	1	2	3	4	5	10	15	20
	(nitrogen input, lbs./acre/year)							
Chicken manure, 1.6% N								
0.90, 0.10, 0.075, 0.05, 0.04, 0.03	222	220	218	217	216	214	212	210
0.90, 0.10, 0.05	222	220	219	218	217	213	209	207
Fresh bovine waste, 3.5% N								
0.75, 0.15, 0.10, 0.075, 0.05, 0.04, 0.03	267	253	246	242	240	231	223	218
0.75, 0.15, 0.10, 0.05	267	253	246	244	241	230	221	215
Dry corral manure, 2.5% N								
0.40, 0.25, 0.06, 0.03	500	312	349	332	326	295	272	255
0.40, 0.25, 0.06	500	312	349	316	308	258	232	218
Dry corral manure, 1.5% N								
0.35, 0.15, 0.10, 0.075, 0.05, 0.04	571	412	367	343	336	291	270	240
0.35, 0.15, 0.10, 0.05	571	412	367	364	344	281	245	225
Dry corral manure, 1.0% N								
0.20, 0.10, 0.075, 0.05, 0.04, 0.03	1000	600	490	475	451	361	300	261
0.20, 0.10, 0.05	1000	600	580	489	437	277	225	208
Liquid sludge, 2.5% N								
0.35, 0.10, 0.06, 0.05, 0.04, 0.03	571	465	427	400	384	331	292	265
0.35, 0.10, 0.05	571	465	437	406	379	290	245	223

*The first decay series presented is meant to represent a slower rate of mineralization of the residual N from each yearly application.

synthetic organic fertilizers is the same as the process with natural organics.

ORGANIC SOIL AMENDMENTS

Those organic amendments most commonly used by home gardeners and horticulturists are digested sawdust, wood shavings, ground bark, leaf molds, peats and composts. Wood residues can be incorporated directly or composted. Because they have a high moisture-holding capacity, they make excellent soil mulches for blueberries, strawberries, fruit trees, ornamentals and garden crops. Since woody tissue is low in plant nutrients, extra nitrogen and phosphate are needed when they are composted or added to soil.

Table 8-3. Average Analysis of Organic Materials

	Nitrogen % N	Phos- phoric Acid % P₂O₅	Potas- sium Oxide % K₂O	Organic Matter % O.M.	Cubic Feet per Ton
Bulky organic materials					
Goat manure	2.77	1.78	2.88	60	70
Dairy manure	0.7	0.30	0.65	30	55
Steer manure	2.0	0.54	1.92	60	70
Horse manure	0.7	0.34	0.52	60	75
Hog manure	1.0	0.75	0.85	30	60
Sheep manure	2.0	1.00	2.50	60	70
Rabbit manure	2.0	1.33	1.20	50	70
Poultry manure	1.6	1.25	1.9	50	50
Poultry manure	1.6	1.25	0.9	50	50
Seaweed (kelp)	0.2	0.1	0.6	80	—
Alfalfa hay	2.5	0.50	2.10	85	—
Alfalfa straw	1.5	0.30	1.50	82	—
Bean straw	1.2	0.25	1.25	82	—
Grain straw	0.6	0.20	1.10	80	—
Cotton gin trash	0.7	0.18	1.19	80	—
Winery pomace (dried)	1 to 2.0	1.5	0.5 to 1.0	80	—
Olive pomace	1.2	0.8	0.5	80	—
Organic concentrates					
Dried blood	13.0	1.5	—	80	—
Fish meal	10.4	5.9	—	80	—
Digested sewage sludge	2.0	3.01	—	50	—
Activated sewage sludge	6.5	3.4	0.3	80	—
Tankage	7.0	8.6	1.5	80	—
Cottonseed meal	6.5	3.0	1.5	80	—
Bat guano	13.0	5.0	2.0	30	—
Bone meal *	*	*	—	—	—
Castor pomace	6.0	2.5-3.0	0.5	80	—

*There is a wide variation in the average percentages found in bone meal. Average found in analysis of 22 samples ran as follows:

Steamed Bone Meal

	Available Phosphoric Acid, %	Insoluble Phosphoric Acid, %	Total Phosphoric Acid, %
Nitrogen, %			
less than 1.00	12-14	14-16	12-14

All organic materials should be purchased on the basis of actual analysis. There is a wide variation in value due to moisture content, type of storage and other conditions. These values are only averages taken from official literature.

Peats may be divided into two main types, according to the kinds of plants from which they were derived. Sphagnum peat (peat moss) is derived from species of the sphagnum plant, the remains of which have accumulated over centuries in bogs in parts of Canada and the northern U.S. Hypnum peat is derived under similar anaerobic conditions from sedges, reeds, mosses or trees.

Sphagnum peats have a low ash content (below 5 percent), are very low in plant nutrients and are very acid (pH values between 3.0 and 4.5). They do have a very high water-holding capacity equal to 15 to 30 times their own weight, but this is cut in half following drying. They are beneficial as mulches for acid-loving plants; otherwise they may need to be neutralized with limestone.

Hypnum peats are composed of more lignin-like substances quite resistant to further decomposition. Their dry matter contains 5 to 40 percent ash. They do not exhibit as high a water-holding capacity as sphagnum peat moss and are less acid.

MAKING COMPOST

Composting has become a popular method of utilizing organic wastes for conversion to beneficial organic soil amendments. Composting can be carried out on any scale from large industrial waste management to backyard gardening. Almost any natural organic product can be composted with proper care—cornstalks, straw, leaves, grass clippings, winery refuse, garbage, etc. The microbes are not selective.

Compost heaps should be built no more than 6 feet high so that air can enter at the bottom of the pile. Residue should be cut less than 6 inches in length. Nitrogen fertilizer should be added at a rate of ¼ pound of nitrogen per cubic foot of dry material.

Moisture content between 50 and 70 percent is desirable. Decomposition is slowed down when the heap is drier; anaerobic conditions exist, particularly at the bottom of the heap, where it is wetter. Water and nitrogen are best applied to the layers as the pile is built up. If the moisture becomes excessive it can be reduced by loosening the pile. Allowing temperatures to build up will hasten the decomposition. High-temperature-activated organisms complete the composting process.

In farm and garden practices, the compost should be turned every three or four days. Under favorable conditions, composting can be complete in three or four weeks. The compost is complete when it has cooled, has a dark color and is crumbly and odorless. Composts are

particularly beneficial for soils low in organic matter and where frequent tillage and complete removal of crops may lead to soil deterioration.

Composts, like other organic amendments, benefit the soil principally by improving soil structure, water penetration and moisture-
holding capacity. They do not contain enough nutrients to satisfy
growing plants unless these have been added as fertilizer prior to
mixing with the soil.

Table 8-4. Typical Analyses for Organic Soil Amendments

Organic Amendment	Moisture Retention	pH	Organic Matter	Ash
	(% total volume)		(%)	(%)
Sphagnum peat moss	60-70	3.2-4.5	95-99	1-5
Hypnum peat moss	55-65	4.4-6.7	70-85	15-30
Reed and sedge peat	50-60	4.5-7.0	85-95	5-15
Woody peat	30-40	3.6-5.5	75-90	10-25
Sawdust	30-40	3.8-8.0	95-99	1-5
Ground bark	30-40	4.0-8.0	90-95	5-10
Compost	20-30	4.0-8.0	80-85	15-20
Leaf mold	20-30	4.0-7.0	50-75	25-50

ORGANICALLY AMENDED SOIL MIXES

These are sometimes referred to as artificial soils and have experienced a large increase in popularity in recent years. They developed on a commercial scale beginning in the mid-1950's as a consequence of an overabundance of wood by-products from logging
operations. With the growth of the container and landscape industry
has come a general shortage of the more desirable amendments,
mainly redwood bark and sawdust. Organically amended soil mixes
have become an integral part of the controlled plant culture. Their
uses extend from growing hothouse vegetables to producing nursery
plants, and building golf greens and athletic fields. Homeowners pur-
chase these mixes for house and patio plants.

There are many reasons for using organically amended soil mixes.
Often they provide a superior medium for producing quality plants.
The use of turf and ornamental plants in roof gardens and above-
ground patios necessitates low-density growing media. Low-density
soils reduce transportation costs.

In choosing organic soil amendments, the first consideration should be to use materials resistant to decomposition. Redwood bark and sawdust, fir bark and sawdust and pine bark and sawdust are the more resistant wood by-products. Redwood bark is only 2.2 percent decomposed, and redwood sawdust is only 5.3 percent decomposed in 160 days. Some pine sawdust, on the other hand, may be over 50 percent decomposed in 160 days.

Moisture retention and pore space are important considerations when choosing various amendments of different particle sizes. Percent moisture retention is usually given as volume percent for purposes of convenience. For instance, if redwood sawdust has a moisture retention capacity of 50 percent, then a 1-gallon can of sawdust holds ½ gallon of water. It is desirable to use products which hold large volumes of water so long as there is sufficient air present. Total porosity and air accessibility are two very important physical properties of any growing medium. Reliable nurserymen are well aware of the media requirements for a wide variety of ornamentals.

Since organic soil amendments do not have high nutrient-supplying capacities, amended soil mixes require abundant quantities of the nutrient elements, particularly nitrogen. Many commercial nursery operations are on a constant feeding program.

Synthetic soils may contain from 50 to 100 percent organic amendments. Typical soil mixes prepared for landscaping, golf greens, nursery containers, bedding plants, cut flower beds and athletic field turf consist of 50 percent organic amendment mixed with fine sand or

Table 8-5. Approximate Root Aeration Requirements of Selected Ornamentals Expressed as Percent Air Space

Very High 20	High 20 - 10	Intermediate 10 - 5	Low 5 - 2
Azalea	African violet	Camellia	Carnation
Fern	Begonia	Chrysanthemum	Conifers
Orchid, epiphytic	Daphne	Gladiolus	Geranium
	Foliage plants	Hydrangea	Ivy
	Gardenia	Lily	Palm
	Gloxinia	Poinsettia	Rose
	Heather		Stocks
	Orchid, terrestrial		Strelitzia
	Podocarpus		Turf
	Rhododendron		
	Snapdragon		

loam soil plus nutritive or pH-adjusting minerals. The organic amendment is normally redwood or fir bark and/or sawdust, usually with nitrogen previously incorporated. Sphagnum peat moss is often a part of the organic component because of its high moisture retention qualities.

Potting soils and soils for landscape planters contain at least 2/3 organic amendment mixed with the same constituents as previously listed. Specialty crops such as azaleas, rhododendrons, heather and orchids are commercially grown in 100 percent organic soil mixes.

Table 8-6. Approximate Number of Containers Filled from One Cubic Yard of Organically Amended Soil Mix*

300 one-gallon cans	1,500 four-inch pots
500 six-inch pots	4,000 three-inch pots
800 five-inch pots	

Three cu yds. per 1,000 sq. ft. will provide 1 in. of depth.

*Source: Used by permission of O. A. Matkin.

Formulations for particular requirements should be obtained from commercial consultants or university horticultural advisors. There are also numerous publications available with more specific information on this subject.

NUTRITIONAL VALUE

Food grown "organically" has no more nutritional value than food grown with chemical fertilizers and pesticides, according to a report of the Institute of Food Technologists. The nutritional value is determined by the genetic inheritance carried in the seed and by the maturity of the crop when it is harvested, the report said. The plants use the nutrients in the soil for their own growth and maturation, and it makes no difference to them where these nutrients come from.

Everything the plant uses, and everything thus passed on to those who eat the plant or its fruit, is broken down into minute molecules, and this process takes place equally well whatever the source of the nutrients.

The IFT summary reported that, contrary to the dictionary definition which states that all food is organic because it is the product of living organisms, plant or animal, "In today's popular jargon, however, the term 'organic' has taken on a new connotation referring to food from soil that has been treated only with animal manure and composted materials."

The report also discussed the recycling of municipal wastes to solve the problems of waste disposal and nutritional deficiencies.

It concluded that it is not feasible now to transport large amounts of wastes from cities to farms, and that such wastes often contain metals that are poisonous to plants.

According to one authority, the Harvard nutritionist Jean Mayer, the so-called "organic" foods may escape pollution by chemicals, but they tend to become the most biologically polluted of all foods.

"Organic fertilizers of animal or human origin are obviously the most likely to contain gastrointestinal parasites," Mayer said.

SUPPLEMENTARY READING

1. *Chemistry of the Soil.* Firman E. Bear. Reinhold Publishing Corp. 1965.
2. *Fundamentals of Horticulture.* J. B. Edmond, A. M. Musser and F. S. Andrews. The Blakiston Company, Inc. 1951.
3. *Horticultural and Agricultural Uses of Sawdust and Soil Amendments.* Technical Bulletin. Paul Johnson and O. A. Matkin. Copyright 1968 by Paul Johnson.
4. *The Nature and Properties of Soils.* Eighth Edition. N. C. Brady. The Macmillan Company. 1974.
5. *Peat and Muck in Agriculture.* M. S. Anderson, S. F. Blake and A. L. Mehring. Circular No. 888, U.S. Department of Agriculture. 1951.
6. *Soil, Yearbook of Agriculture.* U.S. Government Printing Office. 1957.
7. "Using Organic Wastes as Nitrogen Fertilizers." P. F. Pratt, F. E. Broadbent and J. P. Martin. *California Agriculture*, 27:6, 1973.

Chapter 9

SOIL AND TISSUE TESTING

Soil and plant tissue analyses are the farmer's best guide to the wise and efficient use of fertilizers and soil amendments. They are the first step of a high yield, maximum profit, best management practices program. Useful recommendations from soil and tissue tests require accurate sampling and analysis and interpretations based on sound research and judgment. The interpretive guides presented here are those that apply generally in the West. The user is advised to contact the local agricultural extension service, experiment station or qualified industry representatives for recommendations for crops in specific areas. These guides should be used only with data obtained from samples collected and analyzed by the procedures specified.

Each diagnostic technique has advantages as well as limitations. Soil and plant analyses do different jobs and should be used in such a way that they support and supplement each other. Soil analyses are most useful in appraising related nutrient requirements and in evaluating soil pH and salt problems. They have the advantage that they can be completed and used prior to planting the crop.

Plant analyses are particularly useful in determining the nutritional status of permanent, deep-rooted crops such as tree fruits, vine crops and alfalfa, where soil samples of the entire feeding zone are difficult to obtain and interpret. They are also useful in diagnosing the causes of poor crop growth, in evaluating the effectiveness of fertilizer treatments, in following the nutrient status of plants throughout the growing season and in managing quality factors of many crops.

Satisfactory recommendations based on soil or tissue tests depend upon three factors: representative sampling, accurate analysis and proper interpretation of the analytical results.

SAMPLING

The sample must correctly represent the soil, or the crop being sampled and the results yielded can be no better than the sample analyzed. No single set of sampling instructions can apply to all situations.

Fig. 9-1. Taking a soil sample.

The proper method for collecting and handling samples is determined by:

1. The use to be made of the analyses.
2. The pattern and ease of recognition of soil or crop variability.
3. Previous and proposed management practices.

Soils are heterogeneous. They vary in both their horizontal and
their vertical dimensions. Management practices such as leveling,
fertilizing and cropping increase this heterogeneity. It is important
that this variability be recognized and, if practical, measured. An
average test value from a highly variable area may be of little use.

In sampling a field, one should first divide it into relatively homogeneous and manageable units. The divisions can be based on the appearance of the crop or soil surface, management history, drainage, texture, erosion, etc. Then he should collect and make a composite of at least 10, preferably 20, subsamples for the sample from each unit. To diagnose the causes of poor or abnormal growth, he will frequently find samples from adjacent normal and affected areas to be the most useful.

The depth to which samples should be taken is determined by the crop, the proposed use of the analysis and what is already known about the soil profile. Evaluation of soluble salt levels and of levels of nutrients that move freely within the profile requires samples taken to the rooting depth. Subsoil samples are frequently of value in explaining unexpected crop growth patterns resulting from either chemical or physical characteristics of subsoil layers. For many uses, samples taken to plow depth are adequate.

The time to take soil samples is determined by the information desired. Samples for determining fertilizer needs of annual crops should be taken sufficiently in advance of planting to allow time for analysis and return of the results from the laboratory. Samples for diagnosing the cause of poor crop growth or for evaluating salt or sodium hazards are best taken while the problem areas are delineated by crop or other visual differences.

In sampling plant tissues, the crop, its stage of growth, the part of the plant to be analyzed and variation within the field must be considered. The nutritional status of most perennial crops, e.g., tree fruits, citrus, grapes and alfalfa, may be evaluated from samples taken at a single growth stage. For other crops, such as vegetables, cereals and sugar beets, it is better to make samplings at several stages of growth so that impending nutrient deficiencies can be detected and corrected before severe yield reductions occur. Interpretation of these analyses is based on guides established from analyses of specific tissues sampled at specified times. For this reason the plant parts collected must correspond with those of the interpretive guide used. Plants, like the soils on which they are growing, are not uniform over large areas, and for this reason a field to be sampled should first be subdivided into uniform soil and cropping areas. Thirty to 50 plant parts are then collected as the samples from an area.

Before submitting either soil or plant tissue samples for analysis, one should:

1. Obtain containers, instructions and information sheets from the laboratory.

2. Follow instructions carefully in collecting and preparing the samples.
3. Keep accurate records of the areas sampled, fertilizers used, crop yields, etc., for future reference.

ANALYTICAL METHODS

It is necessary to choose the kinds of analyses that are meaningful when the sample is sent to the laboratory since laboratories can analyze for many things. Also, the results of these analyses are useful only insofar as the methods are reliable and the data interpretable. Soil analyses of general use in evaluating levels of available nutrients in western soils include pH (measured in a soil-water paste), $NaHCO_3$ extractable P, exchangeable K (neutral salt extraction), and DTPA-extractable Zn. Detailed methods for these determinations are available from several sources (see Supplementary Reading).

Plant tissue analyses are used to identify inadequate, normal or excessive amounts of several elements, thus procedures for a

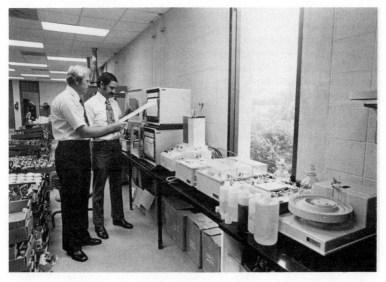

Fig. 9-2. Modern soil and tissue laboratories contain automated equipment such as this Multichannel Technicon Auto Analyser II Continuous Flow Analytical System.

wide range of measurements are required. The most common determinations include the concentration of nitrate (NO_3-N), total or
Kjeldahl N, 2 percent acetic acid–soluble phosphate (PO_4-P), total
P, readily reducible sulfate (SO_4-S) and the total contents of K, Ca,
Mg, S and the micronutrients. Suitable methods for these measurements are available from several sources (see Supplementary Reading).

The results of these analyses are reported as the concentration of
the element in oven-dry soil or plant tissue.

INTERPRETIVE GUIDES

Translation of soil and tissue test data into meaningful management recommendations involves the integration of all available pertinent information. It should be done by experienced agriculturists
who understand the principles of soil fertility and plant nutrition,
and who are familiar with the soils, the fertilizer response data and
crop production in the area.

Soil Tests

Guides to the interpretation of soil tests for available P, K and Zn
are in Table 9-1. The values listed, although generally applicable,
cannot be applied to all situations. Bicarbonate-extractable P is not
a reliable index of phosphate availability in strongly acid or high
organic matter soils. Similarly, interpretations of exchangeable K
levels should take into account the rate of release of nonexchangeable K, subsoil K levels and the influences of levels of other cations.
In spite of their limitations, these soil tests provide growers with
estimates of fertilizer needs of their soils at the time the information
is needed most and is not available from other sources.

Soil pH provides a principal clue in the diagnosis of many soil
problems. Characteristics such as the solubility and availability to
plants of several important nutrient elements, the rates of microbial
reactions and the level of exchangeable sodium are closely related to
the pH of the soil. In strongly acid soils (pH below 5.5), the solubility of aluminum and manganese and other heavy metals is high,
and yields are frequently reduced by toxic levels of one or more of
these elements. At the other end of the scale are the strongly alkaline
soils (pH above 8.5) ; these contain precipitated carbonates (calcareous soils) and, in some cases, appreciable amounts of exchangeable

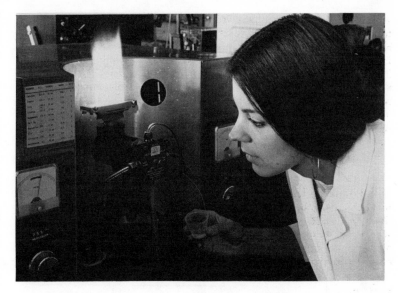

Fig. 9-3. Modern laboratory equipment provides precise analyses.

sodium. Crop growth on these soils may be limited by inadequate supplies of iron (lime-induced chlorosis) or by alkali problems.

Most agricultural soils are in the pH range 5.5 to 8.0. Growth of crops on these soils is influenced by the favorable effects of near-neutral reaction on nitrification, symbiotic nitrogen fixation and the availability of phosphate, sulfur, molybdenum and the macronutrient cations. The optimum pH range for most crops is 6.0 to 7.5 and for legumes, 6.5 to 8.0.

Tissue Tests

Guides to the interpretation of tissue tests for the principal western crops are given in Tables 9-2 through 9-8. Nutrient levels are given in relation to the probability of responses from application of fertilizers supplying the element. Tissue test values at or below the "deficient" levels indicate a high probability of response; those above "sufficient" indicate that adequate supplies are available. Responses may or may not be obtained with tissue levels between "deficient" and "sufficient," depending on production levels and other factors. Concentrations exceeding the "excess" levels may be associated with depressed yield or crop quality. It is very important that samples of the same plant part, taken at the same physiological age, be analyzed if the data presented here are to be used as a guide.

The influence of leaf levels of nitrogen and potassium on yield and quality factors of oranges is presented in Figures 9-4 and 9-5. Increasing the level of a given element in the tree influences some factors favorably and others unfavorably. In most years, returns are maximum when the nitrogen and potassium concentrations are in the optimum ranges. In other years, and in some orchards, adjustments around the optimum ranges may be worthwhile.

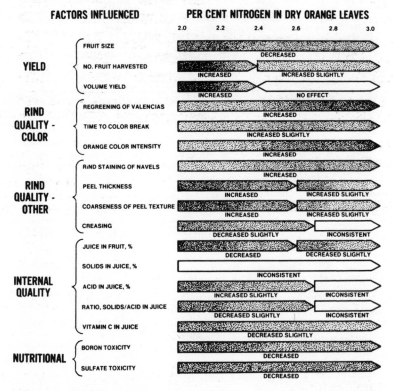

Fig. 9-4. Influence of the percentage of N in 5- to 7-month-old bloom-cycle leaves from nonfruiting shoots upon yield, and rind and fruit quality of oranges (adapted from Embleton *et al.*, 1973).

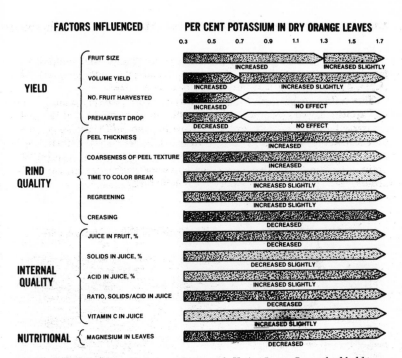

Fig. 9-5. Influence of the percentage of K in 5- to 7-month-old bloom-cycle leaves from nonfruiting shoots upon yield, and rind and fruit quality of oranges (adapted from Embleton, Jones and Platt in *Soil and Plant-Tissue Testing in California*, 1978).

Table 9-1. Generalized Soil Test Interpretive Guide

Most Field and Vegetable Crops	NaHCO₃-P*	Exchangeable K	DTPA Zn
	.... (ppm)		
Highly responsive	<8	<60	<0.5
Probably responsive	8-15	60-100	0.5-1.0
Not responsive	>15	>100	>1.0

Potatoes	NaHCO₃-P		Exchangeable K†	DTPA Zn
	Pac. N.W	S. West		
	.... (ppm)			
Highly responsive	<10	<12	<100	<0.5
Probably responsive	10-20	12-25	100-150	0.5-0.8
Not responsive	>20	>25	>150	>0.8

*For cool season crops, increase these values 30%.
† Exchangeable K values 50 to 100% higher are suggested for Washington and Oregon.

Table 9-2. Plant Tissue Analysis Guide for Western Crops

Crop	Time of Sampling	Plant Part		Nutrient Level* Deficient	Nutrient Level* Sufficient
Alfalfa	One-tenth bloom	Middle half of stem	N	—	—
			P	500	800
			K	0.7	0.8
			SO₄-S	100	200
"	One-tenth bloom	Whole tops	Total P – %	0.17	0.25
			Total K – %	0.80	1.5
			Total S – %	0.17	0.25
Asparagus	Midgrowth of fern	4" tip section of new fern branch	N	100	500
			P	800	1600
			K	1	3
Bean, bush snap	Midgrowth	Petiole of fourth leaf from tip	N	2000	4000
			P	1000	3000
			K	3	5
"	Early bloom	"	N	1000	2000
			P	800	2000
			K	2	4
Broccoli	Midgrowth	Mid-rib of young, mature leaf	N	7000	10000
			P	2500	5000
			K	3	5
"	1st buds	"	N	5000	9000
			P	2000	4000
			K	2	4
Brussels sprouts	Midgrowth	Mid-rib of young, mature leaf	N	5000	9000
			P	2000	3500
			K	3	5
"	Late growth	"	N	2000	4000
			P	1000	3000
			K	2	4
Cabbage	At heading	Mid-rib of wrapper leaf	N	5000	9000
			P	2500	3500
			K	2	4
Cantaloupe	Early growth (short runners)	Petiole of sixth leaf from growing tip	N	8000	12000
			P	2000	4000
			K	4	6
"	Early fruit	Petiole of sixth leaf from growing tip	N	5000	9000
			P	1500	2500
			K	3	5

(Continued)

Table 9-2 (Continued)

Crop	Time of Sampling	Plant Part		Nutrient Level* Deficient	Sufficient
Cantaloupe	First mature fruit	Petiole of sixth leaf from growing tip	N P K	2000 1000 2	4000 2000 4
Carrot	Midgrowth	Petiole of young, mature leaf	N P K	5000 2000 4	10000 4000 6
Cauliflower	Buttoning	Mid-rib of young, mature leaf	N P K	5000 2500 2	9000 3500 4
Celery	Midgrowth	Petiole of newest fully elongated leaf	N P K	5000 2000 4	9000 4000 7
"	Near maturity	"	N P K	4000 2000 3	6000 4000 5
Cucumber (pickling)	Early fruit-set	Petiole of sixth leaf from tip	N P K	5000 1500 3	9000 2500 5
Lettuce	At heading	Mid-rib of wrapper leaf	N P K	4000 2000 2	8000† 4000 4
"	At harvest	"	N P K	3000 1500 1.5	6000† 2500 2.5
Pepper, chili	Early growth	Petiole of young, mature leaf	N P K	5000 2000 4	7000 3000 6
"	Early fruit-set	"	N P K	1000 1500 3	2000 2500 5
Pepper, sweet	Early growth	Petiole of young, mature leaf	N P K	8000 2000 4	12000 4000 6
"	Early fruit-set	"	N P K	3000 1500 3	5000 2500 5

(Continued)

Table 9-2 (Continued)

Crop	Time of Sampling	Plant Part		Nutrient Level* Deficient	Nutrient Level* Sufficient
Potatoes	Early season	Petiole of fourth leaf from growing tip	N P K	8000 1200 9	12000‡ 2000 11
"	Midseason	"	N P K	6000 800 7	9000‡ 1600 9
"	Late season	"	N P K	3000 500 4	6000 1000 6
Rice	Midtillering	Young, mature leaf (the "y" leaf)	N P K	3.0§ 1000 1.2	3.0§ 1000 1.4
"	Maximum tillering	Young, mature leaf (the "y" leaf)	N P K	2.6§ 800 1.0	2.8§ 1000 1.2
"	Panicle initiation	Young, mature leaf (the "y" leaf)	N P K	2.4§ 800 0.8	2.6§ 1000 1.0
Rose clover	Flowering	Leaves	N P K S	—— 1200 0.7 130	—— 1500 1.0 180
Spinach	Midgrowth	Petiole of young, mature leaf	N P K	4000 2000 2	8000 4000 4
Subclover	Third flower	Fully expanded leaves	N P K S	—— 800 0.7 150	—— 1000‖ 1.0 200
Sweet corn	Tasseling	Mid-rib of first leaf above primary ear	N P K	500 500 2	1500 1000 4
Sweet potato	Midgrowth	Petiole of sixth leaf from the growing tip	N P K	1500 1000 3	3500 2000 5
Tomato (canning)	Early bloom	Petiole of fourth leaf from the growing tip	N P K	8000 2000 3	12000 3000 6

(Continued)

Table 9-2 (Continued)

Crop	Time of Sampling	Plant Part		Nutrient Level* Deficient	Sufficient
Tomato (canning)	Fruit 1" diameter	Petiole of fourth leaf from the growing tip	N P K	6000 2000 2	10000 3000 4
"	First color	"	N P K	2000 2000 1	4000 3000 3
Watermelon	Early fruit	Petiole of sixth leaf from the growing tip	N P K	5000 1500 3	9000 2500 5

* Unless otherwise noted, values are: N = NO₃−N, ppm; P = acetic acid-soluble PO₄−P, ppm; K = total K, %; and S = SO₄−S, ppm.
† Nitrate concentrations 30% higher are suggested for winter-grown lettuce in the desert valleys of Arizona and southern California.
‡ Nitrate levels 40 to 60% higher are suggested for potatoes growing in the high valleys of Idaho and Oregon. Approximately 60% of the total P of potato petioles is extracted with 2% acetic acid.
§ Percent "Kjeldahl N."
‖ The sufficient level of P for heavily grazed subcover may be 50% higher.

Table 9-3. Plant Nutrient Levels in Sugar Beets*

Nutrient	Plant Part	Nutrient Range in Which Deficiency Symptoms May Appear	Critical Concentration†
		(ppm)	(ppm)
NO₃–N	Petiole	70-200	1,000
PO₄–P	Petiole	150-400	750
	Blade	250-700	—
Total K			
>1.5% Na	Petiole	2,000-6,000	10,000
	Blade	3,000-6,000	10,000
<1.5% Na	Blade	4,000-5,000	10,000
Total Ca	Petiole	400-1,000	1,000
	Blade	1,000-4,000	5,000
Total Mg	Petiole	100-300	—
	Blade	250-500	—
SO₄–S	Blade	50-200	250
Total Mn	Blade	4-20	10
Total Fe	Blade	20-55	55
Total Zn	Blade	2-13	9
Total B	Blade	12-40	27

*Adapted from Hills, Sailsbery and Ulrich, 1978.
†The critical concentration is that nutrient concentration at which growth of a plant is retarded by 10%.

Table 9-4. Interpretative Guide for Cotton Petiole Analysis*

Time of Sampling	Desirable or "Safe" Levels		
	$NO_3–N$	$PO_4–P$	K
	(ppm)	(ppm)	(%)
First square	15,000-18,000	—	—
First bloom	12,000-18,000	1500-2000	4.0-5.5
Peak bloom	3,000- 7,000	1200-1500	3.0-4.0
First open boll	1,500- 3,500	1000-1200	2.0-3.0
Maturity	Less than 2,000	800-1000	1.0-2.0

*Data are for composition of petioles of the most recent, fully expanded leaf on the main stem (usually the third or fourth leaf from the terminal). From data of D. M. Bassett, A. J. Mac-Kenzie and T. C. Tucker.

Table 9-5. Interpretative Guide for Grape Tissue Analysis*

Nutrient	Deficient	Sufficient	Excess
$NO_3–N$, ppm	<350	600-1200	>2400
P (total), %	<0.15	0.30-0.60	—
K (total), %	<1.0	1.50-2.50	>3.0
Mg (total), %	<0.3	0.5-0.8	>1.0
Zn (total), ppm	<15	25-50	—
B (total), ppm	<25	40-60	>300†
Cl (total), %	—	0.05-0.15	>0.50

*Concentrations of nutrients in Thompson Seedless grape petioles collected opposite the cluster at full-bloom. Adapted from information supplied by J. A. Cook.
† Concentration in leaf-blade tissue.

Table 9-6. Leaf Analysis Guide for Fruit and Nut Trees—Pacific Northwest*

	Nutrient	Level of Nutrient in Leaves†		
		Deficient Below	Satisfactory	Excess Above
Peach	N	2.5%	2.6-3.5%	3.5%
Sweet cherry	N	2.0%	2.0-3.0%	3.0%
Apricot, prune	N	2.0%	2.0-2.5%	3.0%
Filbert, walnut	N	2.0%	2.2-2.8%	2.8%
Apple, pear	N	1.7%	1.9-2.5%	2.5%
Apple, pear	K	0.9%	1.0-1.5%	—
Peach	K	1.0%	1.5-3.5%	—
Sweet cherry	K	1.0%	2.3-2.8%	—

(Continued)

Table 9-6 (Continued)

		Level of Nutrient in Leaves†		
	Nutrient	Deficient Below	Satisfactory	Excess Above
Prune	K	1.0%	1.7-2.7%	—
Filbert, walnut	K	1.0%	1.2-?	—
Apple, pear	B	20 ppm	25-50 ppm	80 ppm
Stone fruits	B	20 ppm	35-80 ppm	100 ppm
All fruits	Zn	10 ppm	17-?	—

*Adapted from Oregon State University FG 23, 24, 25, 26, 34 and 35; and Washington State University FG 28f and 28g.
† Midshoot leaves sampled between July 15 and August 15.

Table 9-7. Leaf Analysis Guide for Fruit and Nut Trees—California*

	N† Adequate	K‡		Na§ Excess	Cl§ Excess	B	
		Deficient	Adequate			Adequate	Excess
	(%)	(%)	(%)	(%)	(%)	(ppm)	
Almond	2.0-2.5	1.0	1.4	0.25	0.3	30-65	85
Apple	2.0-2.4	1.0	1.2	---	0.3	25-70	100
Apricot (ship)	2.0-2.5	2.0	2.5	0.1	0.2	20-70	90
Apricot (can)	2.5-3.0	2.0	2.5	0.1	0.2	20-70	90
Sweet cherry	2.0-3.0	0.9	--	--	--	---	---
Fig	2.0-2.5	0.7	1.0	--	--	---	300
Olive	1.5-2.0	0.4	0.8	0.2	0.5	20-150	185
Nectarine and peach (freestone)	2.4-3.3	1.0	1.2	0.2	0.3	20-80	100
Peach (cling)	2.6-3.5	1.0	1.2	0.2	0.3	20-80	100
Pear	2.3-2.8	0.7	1.0	0.25	0.3	20-70	80
Plum (Japanese)	2.3-2.8	1.0	1.1	0.2	0.3	30-60	80
Prune	2.3-2.8	1.0	1.3	0.2	0.3	30-80	100
Walnut	2.2-3.2	0.9	1.2	0.1	0.3	26-200	300

Adequate levels for all fruit and nut crops: P, 0.1-0.3%, Cu, over 4 ppm: Mn, over 20 ppm; Zn, over 15 ppm.

*Leaves are July samples from nonfruiting spurs on spur-bearing trees, fully expanded basal shoot leaves on peaches and olives, and terminal leaflets on walnut. Adapted from information of K. Uriu, J. Beutel and O. Lilleland.
† Percentage N in August and September samples can be 0.2-0.3% lower than July samples and still be equivalent. Nitrogen levels higher than underlined values will adversely affect fruit quality and tree growth. Maximum N for Blenheims should be 3.0% and for Tiltons, 3.5%.
‡ Potassium levels between deficient and adequate are considered "low" and may cause reduced fruit sizes in some years. Potassium fertilizer applications are recommended for deficient orchards but test applications only for "low" K orchards.
§ Excess Na or Cl causes reduced growth at the levels shown. Leaf burn may or may not occur when levels are higher. Salinity problems can be confirmed with soil or root analysis.

Table 9-8. Leaf Analysis Guide for Diagnosing Nutrient Status of Mature Valencia and Naval Orange Trees *

Element		Deficient Below	Ranges† Optimum	Excess Above
N	%	2.2	2.4 to 2.6	2.8
P	%	0.09	0.12 to 0.16	0.3
K‡	%	0.40	0.70 to 1.09	2.3?
Ca	%	1.6?	3.0 to 5.5	7.0?
Mg	%	0.16	0.26 to 0.6	1.2?
S	%	0.14	0.2 to 0.3	0.6
B	ppm	21	31 to 100	260
Fe§	ppm	36	60 to 120	250?
Mn§	ppm	16	25 to 200	1,000?
Zn§	ppm	16	25 to 100	300
Cu	ppm	3.6	5 to 16	22?
Mo	ppm	0.06	0.10 to 3.0	100?
Cl	%	?	<0.3	0.7
Na	%		<0.16	0.25
Li	ppm		<3	35?
As	ppm		<1	5
F	ppm		<1 to 20	100

*With the exception of N values, this guide can be applied to grapefruit, lemon and probably other commercial citrus varieties. Data of Embleton, Jones and Platt in *Soil and Plant-Tissue Testing in California,* 1978.

†Based on concentration of elements in 5- to 7-month-old, spring-cycle leaves from nonfruiting terminals. Leaves selected for analysis should be free of obvious tipburn, insect or disease injury, mechanical damage, etc., and be from trees that are not visibly affected by disease or other injury.

‡Potassium ranges are for effects on number of fruits per tree.

§Leaves that have been sprayed with Fe, Mn or Zn materials may analyze high in these elements, but those of the next growth cycle may have values in the deficient range.

Other Tests

Other tests used include appraisals of salinity and alkali conditions; water quality evaluations; soil levels of other nutrient elements and organic matter; and soil moisture, texture, density and permeability to air and water. Methods of analysis and interpretive criteria for these determinations are presented in the Supplementary Reading.

SUPPLEMENTARY READING

1. *Analytical Methods for Use in Plant Analysis.* C. M. Johnson and A. Ulrich. California Agr. Expt. Bul. 766. 1959.

2. *Diagnosis and Improvement of Saline and Alkali Soils.* USDA Agricultural Handbook No. 60. 1954.
3. *Diagnostic Criteria for Plants and Soils.* H. D. Chapman, Editor. University of California, Division of Agricultural Sciences. 1966.
4. *Fertilizer Technology and Use*, Second Edition. R. A. Olsen, T. J. Army, J. J. Hanway and V. J. Kilmer. Soil Sci. Soc. of America. 1971.
5. *Hunger Signs in Crops*, Third Edition. H. B. Sprague, Editor. David McKay Co., Inc. 1964.
6. "Leaf Analysis as a Diagnostic Tool and Guide to Fertilization." T. W. Embleton, W. W. Jones, C. K. Labanauskas and Walter Reuther. *The Citrus Industry*, Vol. III (Revised). University of California, Division of Agricultural Sciences. 1973.
7. *Methods of Soil Analysis*, Part II. A. L. Page, Editor. Amer. Soc. Agronomy. 1980.
8. *Soil and Plant-Tissue Testing in California.* H. M. Reisenauer, Editor. University of California, Division of Agricultural Sciences. 1978.
9. *Soil Testing and Plant Analysis.* L. Walsh and J. D. Beaton, Editors. Soil Sci. Soc. of America. 1973.
10. *Soil Testing Procedures for California.* CFA-SIC Publication. 1980.
11. *Sugarbeet Fertilization.* F. J. Hills, R. Sailsbery and A. Ulrich. University of California, Division of Agricultural Sciences. Bull. 1891. 1978.

Chapter 10

METHODS OF APPLYING FERTILIZER

Fertilizers are used to supply nutrients that are not present in the soil in amounts necessary to meet the needs of the growing crop. When choosing the methods of application, growers should consider the following:

1. The rooting characteristics of the crop to be planted.
2. The crop's demand for various nutrients at different stages of growth.
3. The physical and chemical characteristics of the soil.
4. The physical and chemical characteristics of the fertilizer materials to be applied.
5. The availability of moisture.
6. The kind or irrigation systems used if irrigation is the only, or major, source of water.

Crop production in the western states often requires multiple applications of fertilizer materials. Several methods of applying fertilizer may be employed. For example, lettuce fields may receive a preplant broadcast application of nitrogen and phosphorus fertilizer. Sufficient quantities of phosphorus, a non-mobile nutrient, can be used in this preplant application to satisfy the crop's needs for the entire season. As the crop grows, additional nitrogen is directly injected into the beds near the row, water-run, or perhaps both.

Fertilizers added to soils undergo transformations that may change their availability. The methods of application are directly related to the crop's utilization of the nutrients and the changes the nutrients undergo in the soil. The application methods employed should be as economic, accurate and efficient as possible.

This chapter discusses different methods of fertilizer application and the equipment used in applying fertilizer. It suggests methods of adjusting the equipment for accurate applications and ways of overcoming some problems related to fertilizer applications.

PREPLANT APPLICATIONS

Broadcast

The broadcast method of applying fertilizers consists of uniformly distributing dry or liquid materials over the soil surface.

Drop spreader—The simplest applicator for dry fertilizer is the single unit fertilizer spreader consisting of an inverted triangle-shaped hopper. The hopper is mounted between two wheels and is usually pulled by a tractor or pickup truck. Its small size limits the fertilizer load and restricts its use to small fields or to areas where large application equipment cannot enter.

Patterned after the simple fertilizer spreader is a folding unit that holds a large amount of fertilizer and spreads a wider swath (Figure 10-1). The unit folds forward for easy transport to the field. When properly adjusted or calibrated, very accurate application rates can be obtained. The unit is pulled by a tractor or truck.

Pull-type spreader—Small bulk spreaders are available, each consisting of a bin mounted on a two- or four-wheeled trailer frame and pulled by a tractor or truck (Figure 10-2). These spreaders can be pulled across the field at speeds up to 20 mph. The fertilizer swath is

Fig. 10-1. A dual-unit fertilizer drill, folded for transport. The front end is attached to the draw bar and raised off the ground.

Fig. 10-2. A tandem bulk fertilizer spreader that holds up to 10 tons or dry material.

usually 20 to 40 feet wide. The operator must exercise care to prevent skips or excessive overlaps of fertilizer.

Self-propelled spreader—The bulk fertilizer spreader consists of a large bin mounted on a large truck or a special three- or four-wheeled vehicle equipped with flotation tires (Figure 10-3). The bin can hold from 7 to 10 tons of dry fertilizer material.

The fertilizer is spread in a swath 30 to 60 feet wide by means of one or two rapidly spinning horizontal discs at the rear of the bin. Bulk spreading equipment can travel across fields at speeds up to 35 mph. Proper adjustment of the metering device and speed, and care on the part of the operator to avoid misses, gaps between or overlaps of the swath, permits a reasonably accurate, fast and economical application of fertilizer.

Liquid spreader—The basic requirements for a liquid fertilizer broadcast applicator consist of a tank, pressure gauge and regulator, pump, pipes, hoses, fittings, nozzles and a boom. The applicator

can be mounted on a truck, flotation vehicle, trailer or three-point
hitch mounted on a tractor. The speed of application is determined
by the rate of flow.

Injection

Injection refers to placing fertilizers below the soil surface which
is accomplished by using tool bar-mounted knives or shank openers.
Drop pipes for liquid fertilizers, or flexible tubes for dry fertilizers,
deliver fertilizers into the channels made by the opening tools. All
fertilizers that can be broadcast on the soil surface can also be injected. Because certain fertilizers are subject to nitrogen losses when
they are applied to the soil surface, many growers choose to inject
them below the soil surface.

Injection is also an excellent method of putting immobile nutrients
into the crop root zone. If soils are subject to erosion, injection helps

Fig. 10-3. A three-wheeled 10-ton bulk fertilizer spreader equipped with
flotation tires.

to prevent nutrient losses from occurring when the soil particles are carried away by the wind or by sudden intensive rains. Fertilizers are usually injected after plowing or disking or when furrowing out. The soil is loose and crop residues will be dispersed.

AT PLANTING APPLICATIONS

Injection of liquid fertilizers requires a tank, a pump, a pressure gauge and regulator, hoses, fittings, tool bars and injection knives or shanks. It should be noted that special high pressure equipment is required for injecting anhydrous ammonia. Because the power requirement for injecting fertilizers is great, a tractor must be used. Tanks for liquid fertilizers, particularly anhydrous ammonia, may be mounted on a tool bar or saddle-mounted, front-end-mounted or mounted on a trailer and towed behind the tractor.

Equipment for injecting dry fertilizers is less elaborate. Usually a large hopper is mounted on the rear tool bar. The metering wheels on the hopper are driven by a chain drive from the tractor wheel, a ground drive unit or a hydraulic motor. Drop tubes connecting the hopper to the injection shanks allow the placement of the fertilizer in the soil.

Anhydrous ammonia and aqua ammonia are usually applied in bands 12 inches or more apart. Modern high-powered tractors permit use of tool bars 24 feet wide, or greater, for the application of these fertilizers. Liquid and dry fertilizers are usually applied preplant at the time furrows and beds are formed. The injection shanks are placed above the ground level about halfway between the point of the lister shovel and the top of the bed. The fertilizer is metered into the soil as it is thrown into the bed. Bed shapers, either included as part of the total injection-bedding operation, or pulled through the field after injecting is finished, leave the field ready for planting.

Placement with Seed

This method of application consists of placing the fertilizer directly with the seed. The fertilizer hopper is attached to the grain drill, and the fertilizer is metered out according to predetermined rates. Newer grain drills have fertilizer hoppers that can hold up to 1,800 pounds of dry fertilizer. This new equipment lends itself to the use of bulk fertilizer field tenders, eliminating the time-consuming job of breaking bags. Practical liquid fertilizer applicators for grain drills have not yet appeared on the farm equipment market.

Small grain crops such as wheat, barley and oats have been observed to respond better to fertilizers placed with the seed, particularly when low rates of fertilizers that contain phosphorus are used. The response is less obvious at higher fertilizer rates and on soils that have a low phosphorus-fixing capacity. Growers should avoid using too high of rates of fertilizer, particularly with nitrogen, since germination can be affected.

Band Placement

Ths method of application consists of placing fertilizer to the side and/or below the seed. Applicators for banding dry fertilizers on row crops are usually mounted on the same tool bar or bed sled as the seeder and are powered by the same drive that powers the seeder. A tube connected to the fertilizer hopper delivers the fertilizer to a furrow opened by a shoe or disc. The tank for liquid fertilizers may be on the tool bar, or saddle mounted on the tractor. The injection shank of liquid applicators is usually either attached to the seeder or set to inject the fertilizer ahead of the seeder. The band applicator can be set to place the fertilizer in bands at any depth or position relative to the seed. Depending on the crop and soil type, the band may be placed to the side and below the seed, or it may be placed directly below the seed.

A relatively new and related method of supplying some of the plant nutrients is pop-up fertilizer application. With this method, a small amount of fertilizer is applied directly with the seed. Pop-up fertilizers may be liquid or dry. The overall effect of pop-up application is to hasten growth once the plant emerges. Pop-up fertilizer application may, in some instances, be an effective substitute for band application.

POST EMERGENCE APPLICATIONS

Side Dressing

This method of fertilization refers to the placement of fertilizers beside the crop rows. These applications may be made at the same time the rows are cultivated. Depending upon the soil type, effects of irrigation and the crop, one or several side-dressed applications may be desirable.

Both liquid and dry fertilizers may be side dressed. The kind of fertilizer being side dressed determines its placement relative to the crop row. Side-banded phosphorus and potassium fertilizers must be placed close enough to the row to be available to the crop's roots. The placement of most nitrogen fertilizers is not as critical because of the mobility of nitrogen fertilizers in the soil. Placement of fertilizers containing nitrate should not be too shallow alongside the row because nitrate will move upward and away from the root zone as irrigation water subs to the row and bed surface. Placement of anhydrous ammonia or aqua ammonia should be farther out and deeper than the seed level or plant roots. Plants are sensitive to high concentrations of ammonia. More distance between the crop row and the side-dressed band of anhydrous ammonia or aqua ammonia allows space for these materials to diffuse into the soil, thereby reducing the ammonia concentration.

Equipment for side dressing dry fertilizer usually consists of two large hoppers mounted on tool bars on either side of a tractor, ahead of the operator or one or more hoppers mounted on a three-point hitch tool bar behind the operator. The metering device is tractor, ground or power driven. Flexible tubing connects the hopper to spouts mounted ahead of, between or behind discs or shovels. The desired amount of fertilizer is metered into a small furrow and is covered by the cultivating unit. Liquid side-dressed fertilizer tanks are usually saddle mounted and are connected to a pump and flow control. Hoses from the control are connected to the injection shanks or spouts mounted ahead of, between or behind discs or shovels. Fertilizer flows into the shank channels or small furrows and is covered by the cultivating unit.

Top Dressing

This method of application consists of placing fertilizers on the soil surface after crop emergence. Field crops such as small grains, pastures, rangeland and alfalfa can be top dressed once or several times during the growing season. The same equipment used for preplant broadcast applications of dry and liquid fertilizer may be used for top dressing. Vehicles used for top-dressed applications should be equipped with flotation tires to minimize crop damage and soil compaction. Aerial applications may be used to eliminate damage from equipment. Aerial top-dressed applications are made on rice because ordinary top-dressing equipment cannot enter the flooded fields.

WATER-RUN APPLICATIONS

Water-run fertilizer applications are not as accurate as equipment-applied applications. Savings in time, labor, equipment costs and

fuel costs and reasonably good results may justify the practice.
Water-run applications of fertilizer may be preplant or post-emergence
applications. Both liquid and dry fertilizer materials may be used.
The plant nutrients should not be introduced into the system at the
initiation of the irrigation set. Best results are obtained when the
fertilizer application is gauged to enter the system toward the middle
of the set and to terminate shortly before the set is completed. Application of fertilizers through the irrigation system in this manner prevents the nutrients from being leached beyond the reach of the roots
or from lying near the surface, inaccessible to the crop.

System and Fertilizer

Open systems—These systems include lined and unlined open
ditches and gated pipes that are used for furrow and flood irrigation
methods. All liquid and dry fertilizers, and liquid or slowly soluble
soil amendments, can be applied through open systems. The application of fertilizers through these systems is usually limited to nitrogen fertilizers. The effectiveness of all nitrogen fertilizers applied
through open systems may not be the same, however. Under hot desert

Fig. 10-4. Application of anhydrous ammonia directly into irrigation water.

conditions, nitrogen losses are likely to occur with anhydrous ammonia or aqua ammonia. A more stable nitrogen solution such as an ammonium nitrate solution or a urea–ammonium nitrate solution may be a better choice of fertilizer.

Profitable crop responses have been reported with the use of phosphorus and potassium fertilizers applied in open systems. Preplant applications of phosphorus and potassium fertilizers may produce profitable responses if the soil is dry and cracked enough to allow the fertilizers to penetrate into the soil profile. There is some question as to the desirability of applying these fertilizers postplant. Since they are less mobile than nitrogen, particularly phosphorus fertilizers, they cannot readily move into the root zone. Because these nutrients are less mobile, water-run applications are not always the best method for applying them.

Liquid soil amendments such as ammonium polysulfide and calcium polysulfide can be applied directly into open systems through metering applicators. Gypsum can be added to the water through weir boxes by diverting part of the stream through a pile of gypsum (see Figure 7-4). About one ton of high grade gypsum can be dissolved in an acre-foot of water. Sulfuric acid may be applied as a soil amendment; however, it is highly corrosive and will attack concrete liners and metal gates in the irrigation system.

Equipment for applying dry fertilizer materials through open systems may be inefficient or cumbersome. It may consist of a field worker standing by the irrigation canal pouring a coffee can of fertilizer into the water every so often. It can also be an elaborate platform along side or straddling the canal, complete with a hopper and a power- or water-driven mechanism to meter the fertilizer into the water. Liquid fertilizers, on the other hand, present fewer problems with respect to the ease of their introduction into irrigation systems.

Application of fertilizer solutions through open systems requires less sophisticated equipment than that needed for applying anhydrous ammonia. Tanks are set up by the weir box or check drops or along side the irrigation ditch. The tank is connected to a float box that meters the material into the water. The float box is usually a hard rubber box about the size of an automobile battery casing. The valve is activated by a toilet tank float. A more elaborate setup is required for applying anhydrous ammonia through the irrigation water. Equipment necessary for injecting anhydrous ammonia into open irrigation systems consists of a liquid-out valve, a tank, a quick-connect coupling, a line strainer, a flow control pressure regulator,

Fig. 10-5. Water-running a nitrogen solution. The fertilizer is being metered into the irrigation ditch by a float box.

a back-flow check valve, an orifice, a length of hose and a spreader tube at the point where the ammonia enters the water.

Sometimes anhydrous ammonia is metered into the water by means of a fixed orifice flow control. Regardless of the metering system selected, the proper size orifice to be used is determined from the pressure regulator and orifice charts provided by the manufacturer. When injecting anhydrous ammonia into the irrigation ditch, the grower should put the tank about 50 feet upstream from the outlets into the field to allow the ammonia to mix well with the water.

Anhydrous ammonia can be directly injected into underground lines that distribute large volumes of water throughout the field. The only difference between this method of application and the one just described is that the spreader tube is inserted through a tap in the irrigation line. Either a flow control regulator with various orifices or fixed orifice flow control can be used. In determining regulator gauge readings, the grower must obtain the waterline pressure and add to the reading obtained from the pressure regulator and orifice charts.

Closed systems—These systems include sprinkler, spitter, trickle, drip and dual-wall tubing systems. Sprinkler systems usually have all metal plumbing, while the others employ a considerable amount of plastic tubing and fittings.

Not all dry and liquid fertilizers can be applied to closed systems. Of the soil amendments, only sulfuric acid can be considered for application through closed systems. Because it is highly corrosive, its application is limited to systems constructed mainly of plastics. Filters are desirable when one is applying fertilizers through closed irrigation systems. All fertilizers should be introduced into the system at a point well ahead of the filters.

Dry nitrogen fertilizers can be applied through closed systems provided: (1) they do not contain coating materials that will most certainly plug the filter or system and (2) they are predissolved before being introduced into the system. Local fertilizer dealers should be consulted for information regarding coating materials on dry nitrogen fertilizers.

Since most dry phosphorus fertilizers are not completely water soluble and may contain some insoluble impurities, it is not recommended that they be applied through closed systems.

Dry potassium fertilizers may be applied through closed systems provided they are predissolved before being introduced into the system. Because potassium is relatively immobile in the soil, it is doubtful that economic responses can be obtained by applying this nutrient through sprinkler systems. Permanently placed spitters, spaghetti tubing in trickle systems, emitters in drip systems and surface or subsurface dual-wall systems are more likely to permit a potassium response. Permanently placed emission points permit a localized concentration of the potassium fertilizer. Potassium moves downward into the soil by mass flow.

Liquid nitrogen fertilizers, except anhydrous ammonia and aqua ammonia, are very adaptable to all closed irrigation systems. Anhydrous ammonia and aqua ammonia have a tendency to volatilize at the discharge point, thus resulting in high nitrogen losses. Moreover, these two fertilizers can cause plugging if the irrigation water has a high calcium content. Whether or not the closed system is constructed with a sophisticated filtering system, in-line filters between the field storage tank and the injection pump should be used because loose scales in the tank, rather than the fertilizers, are usually responsible for most plugging problems.

There are several liquid phosphorus fertilizers that can be used in closed systems. These should be metered into the line by means of a tap and a high pressure, low volume pump. It is quite unlikely that the application of phosphorus fertilizers through sprinkler systems will result in a direct economic response. Phosphorus is immobile and will remain on the soil surface, inaccessible to most crop roots until some physical force moves it into the root zone. However, factors other than mobility influence the probability of response with other closed irrigation systems. For example, crop response to acidified phosphorus solutions applied through permanently placed, low volume irrigation systems is possible because the nutrient is locally concentrated. Phosphorus moves downward through the soil by redissolution.

Liquid ammonium orthophosphate and ammonium polyphosphate fertilizers react with calcium in the water to produce insoluble calcium phosphate compounds that may plug the system. The higher the calcium content of the water, the faster the plugging rate. This problem can be eliminated by introducing an acid into the system at a rate that maintains the solution at a pH of about 5. Phosphoric acid can be used as a phosphorus fertilizer without plugging problems by applying a rate which provides a slightly acid solution.

Special tanks are available on the market for dissolving dry fertilizers for application through closed systems. For liquid fertilizers, tanks can be set up at the irrigation pump site or taken to the field being irrigated, depending upon the application system being used. Fertilizers are injected directly into the irrigation lines by means of a tap and a high pressure, low volume pump (Figure 10-4). Various kinds of injection pumps are also available: Some are electric or gasoline motor driven, while others are driven by in-line impellers in the irrigation line. All can be accurately calibrated. Growers using municipal water sources may wish to consult water authorities to determine if a check valve may be required to prevent fertilizer materials from being introduced into the municipal water system.

Water Quality and Fertilizers

When the irrigation water has a high calcium content, the addition of anhydrous ammonia or liquid ammonium polyphosphate fertilizers may increase the sodium hazard of the irrigation water (see Chapter 2). Anhydrous ammonia causes the calcium in the water to drop out as solid lime particles. This allows dissolved sodium to attach itself to the soil particles, thus reducing the water intake capacity of the soil. This effect may be reversed with subsequent irrigations during which no anhydrous ammonia is applied. Ammonium poylphosphate fertilizer can have the same detrimental effect on the water quality by dropping out the calcium as solid calcium pyrophosphate particles. Dissolved sodium is then free to become attached to the soil particles. Because ammonium polyphosphate fertilizers are applied at rates considerably lower than those of anhydrous ammonia, the adverse effect may not be as serious. Growers should have their irrigation water analyzed to determine whether or not these fertilizers can be water-run applied.

FOLIAR APPLICATIONS

Foliar nutrition involves the absorption of nutrients by all above-ground parts of plants. Foliar feeding is used in situations where a quick response is required or where soil-applied nutrients have been ineffective. The following nutrients have been successfully used on a variety of crops: nitrogen, phosphorus, potassium, calcium, magnesium, sulfur, manganese, copper, boron, molybdenum, iron and zinc. Growers interested in foliar feeding should seek the counsel of farm advisors, extension specialists or industrial agronomists regarding the crop to be treated, the nutrients that can be applied and rates and methods of application.

Nutrients can be applied to plant foliage as sprays or through overhead sprinkler systems. The spray method is often preferred to the irrigation method because of more uniform distribution. In many situations, spray application is the only method that can be used. Equipment for applying pesticide sprays is usually used to apply nutrients.

Foliar feeding by aerial application of nutrients to crops such as grain sorghum, corn or small grains is a satisfactory method of application. It is probably the best method to use when ground equipment cannot enter the field or where very large acreages must be covered quickly.

Ground spray equipment used for foliar feeding is usually of the high pressure, low gallonage type, designed to distribute the spray materials uniformly on the foliage and keep water volume to a minimum. The spray may be applied through single- or multiple-nozzle hand guns; multiple-nozzle booms; or by multiple-nozzle, oscillating or stationary cyclone-type orchard sprayers. The response on crops may be affected by droplet size; therefore, the spray used must be regulated by adjusting pressure and selecting the proper nozzles and discs. While the concentration of a nutrient in a foliar application solution varies widely with respect to the nutrient in question and the crop it is being applied to, the concentration is generally less than 1 to 2 percent to prevent injury to the foliage. The addition of an adjuvant to the spray is recommended for better coverage.

With respect to the use of urea in foliar feeding, growers should use material with a low biuret content on many crops. Biuret is a compound formed during the manufacture of urea. It is toxic to some plants and can be especially damaging when applied to plant foliage. Plant scientists generally agree that where sizable applications of urea are made, biuret should not exceed about 0.25 percent under Southwest climatic conditions.

Application of nutrients for foliar feeding through overhead irrigation systems follows some of the same principles observed for applying nutrients to the soil through irrigation. The exceptions, however, are:

1. The concentration of the material being applied to the foliage must be considerably lower than when applied to the soil.
2. The nutrient must be introduced into the system toward the end of the irrigation set so that the nutrients remain on the foliage when the system is shut down.

Whether applied through aerial or ground equipment or overhead sprinkler systems, most crops respond better to foliar feeding when the nutrients are applied during the morning hours.

Somewhat related to foliar feeding is the application of nutrients to vines and fruit and nut trees during the latter part of the dormant season. Methods of application of the nutrients are the same as those for foliar feeding. However, the nutrients may be applied at concentrations somewhat higher than those normally used for foliar feeding.

CALIBRATION OF APPLICATION EQUIPMENT

The rate-of-delivery settings on fertilizer applicators may be pre-determined by the equipment manufacturer. Charts showing the settings or orifice sizes for rates of application according to the kinds of fertilizer are usually affixed to the equipment or are given in the operator's manual. The local fertilizer dealer should be consulted for further information.

FERTILIZER-PESTICIDE MIXTURES

Fertilizer-pesticide mixtures are used to fertilize crops and control soil-borne insects, diseases, nematodes and weeds. There are more difficulties in making dry fertilizer-pesticide mixtures than there are in formulating liquid fertilizer-pesticide mixtures. For example, when making fertilizer-pesticide mixes, the fertilizer must be impregnated with the insecticide because the granular form of pesticide mixed with fertilizer will segregate. Even after successful impregnation, the mixture must be applied within a short time or the pesticide begins to deteriorate. Storing the mixture in airtight containers may retard deterioration, but the cost of the containers and handling may erode any economic gains. A uniform mixture of dry materials with pesticides is difficult to obtain because of the small amount of pesticide added to a ton of fertilizer.

Pesticides have been successfully mixed and applied with both liquid mix fertilizers and liquid nitrogen solutions. Growers are encouraged to consult with their farm advisors, extension specialists and industrial agronomists and entomologists for complete details regarding compatibility, stability and use of pesticide-fertilizer mixtures in crop production.

Even though compatibility and stability factors may present no problems, the nature of the job to be accomplished and the grower's crop management practices may be more influential in making the decision to use fertilizer-pesticide mixtures. For example, it would not be profitable to apply a herbicide with a phosphate fertilizer when the herbicide is to be lightly incorporated. The phosphate fertilizer would be inaccessible to the crop roots. Furthermore, there is the question that arises regarding the anticipated responses to fertilizer-pesticide application. If the fertilizer solution is being used merely as a carrier, as it might be in aerial applications, while the pesticide may be applied at the correct rate, the amount of fertilizer that will be applied may be inadequate to produce profitable crop responses.

The application of pesticides separately generally produces better results. Control of application timing, placement and rate are more precise with this practice.

Apart from the agronomic standpoint, there is the aspect of entomological and regulatory problems with the use of fertilizer-pesticide mixtures. Recent legislation, for example, provides that only qualified licensed individuals can formulate, sell, recommend and apply pesticides. Once a pesticide is added to a fertilizer, the mixture takes on a completely different set of legal ramifications regarding its use. It would be advisable for growers to consult with qualified and licensed technologists before mixing and/or applying fertilizer-pesticide mixtures.

SUPPLEMENTARY READING

1. *Advances in Production and Utilization of Quality Cotton: Principles and Practices.* Fred C. Elliot, Marvin Hoover and Walter K. Porter, Jr. Iowa State University Press. 1968.

2. *Advances in Sugarbeet Production: Principles and Practices.* Russell T. Johnson, John T. Alexander, George E. Rush and George R. Hawkes. Iowa State University Press. 1969.

3. *Agricultural Anhydrous Ammonia: Technology and Use.* M. H. McVickar, W. P. Martin, I. E. Miles and H. H. Tucker. Soil Sci. Soc. of America. 1966.

4. *Changing Patterns in Fertilizer Use.* L. B. Nelson, M. H. McVickar, R. D. Munson, L. F. Seatz, S. L. Tisdale and W. C. White. Soil Sci. Soc. of America. 1968.

5. *The Citrus Industry,* Volume II. Leon Dexter Batchelor, Walter Reuther and Herbert John Webber. University of California. 1968.

6. *The Citrus Industry,* Volume III. Walter Reuther. University of California. 1973.

7. *Fertilizer Technology and Usage.* M. H. McVickar, G. L. Bridger and L. B. Nelson. Soil Sci. Soc. of America. 1963.

8. *Fertilizer Technology and Use,* Second Edition. R. A. Olson, T. J. Army, J. J. Hanway and V. J. Kilmer. Soil Sci. Soc. of America. 1971.

9. *Soil Fertility and Fertilizers,* Third Edition. S. L. Tisdale and W. L. Nelson. The Macmillan Company. 1974.

10. *Using Commercial Fertilizers,* Fourth Edition. M. H. McVickar and W. W. Walker. The Interstate Printers & Publishers, Inc. 1978.

Chapter 11

GROWING PLANTS IN SOLUTION CULTURE

The term *hydroponics* refers to the growing of plants with their roots immersed continuously or intermittently in a water solution containing the essential mineral nutrients. The rooting medium may be the solution itself or, quite commonly, a pea gravel or coarse sand used to give support to the plant roots while the solution is trickled through or flushed periodically to provide water and nutrients.

Scientists commonly use this method of growing plants in their research work. Food production for military forces was accomplished in this manner during World War II in remote areas where conditions were unfavorable for normal farming with soil. For many years, commercial operations have utilized hydroponics to grow vegetables. Since there is quite a wide interest in this technique, the following discussion is included to serve as a general guide.

SMALL SCALE HYDROPONICS

Those who wish to try hydroponics on a small scale can do so with a fairly small investment in equipment. A one- or two-quart glass jar or crock may be used as the growing vessel. Many attractive containers may also be selected. It is important that the container exclude light, or algae growth will be a problem. Glass or other transparent containers coated with black paint or covered with aluminum foil or some other coating will exclude light. Galvanized containers or those that rust should not be used unless they are lined with polyethylene or similar plastic. A plastic food bag works very well for this purpose.

Put the nutrient solution into the container to within 1 to 1½ inches from the top. Fit the top of the container with a large cork or suitable stopper with a ¾-inch hole in the center. Suspend the seedling or cutting with its roots in the solution. Place some non-absorbent cotton or glass wool around the stem for support and to allow air to circulate. It may be advantageous to use a small aquarium pump to keep the solution aerated. Many plants can be grown in the same jar if a larger container with several holes is used.

For larger scale water culture, use a wooden, concrete, glass or other suitable tank about 6 inches deep, 2 or 3 feet wide and any desired length. Polyethylene or a nontoxic asphalt paint can be used to line the tank. Use a board or other adaptable covering for supporting the plants or a wire netting over which is placed moss, excelsior or a similar material. Use of an air pump is again desirable.

The level of the nutrient solution will drop as plants extract water and as a result of evaporation. Add solution as needed to keep the roots submerged. Every two or three weeks discard the old solution and replace it immediately with a freshly made solution.

Another hydroponic system uses clean, acid washed and rinsed sand in a flower pot or other suitable container with a drainage hole. The seeds or cuttings are grown in the pot and watered daily with the solution, which is allowed to drain into a reservoir. This solution is reapplied daily or as needed and, after two or three weeks, discarded. Just before changing the solution, irrigate plants with fresh water to flush out any accumulated salts.

A variation of this method is to use a small pump, hooked to a time clock, that periodically circulates the solution into the pot. Or use a reservoir of solution which is placed at a higher level than the container. Let the solution drip slowly through a capillary tube using another reservoir to collect the draining solution.

Most water in the home is satisfactory unless it is treated with a water softener. This places too much sodium in the system. It is much better to use hard water that contains calcium and magnesium. Distilled or deionized water may be used but this will increase the need to carefully watch for micronutrient deficiencies, since ordinary water often contains some zinc, copper and other nutrients.

FORMULA FOR SMALL SCALE SYSTEM

The formula shown in Table 11-1 is based on work done by Professor Hoagland at the University of California. The chemicals can be obtained from chemical supply houses, drug stores and garden supply dealers. Technical and fertilizer grade salts are adequate unless nutrient deficiency experimentation is being done. In this case, chemically pure salts, distilled or deionized water and plastic-lined pots will be needed.

Dissolve the potassium phosphate first and then add the other chemicals in the order given. Dissolve each one before adding the next. Warm water will help get them in solution.

Essential micronutrients (also called trace elements) must be added. It is best to make stock solutions of these and then add the

Table 11-1. Components of the Hoagland Solution

Salt	Nutrient(s)	Amount for 25 gallons		
		(gm)	(oz.)	(tbsp.)
Potassium phosphate (monobasic salt)	Phosphorus Potassium	14	½	1
Potassium nitrate (fertilizer grade)	Potassium Nitrogen	57	2	4
Calcium nitrate (fertilizer grade)	Nitrogen Calcium	85	3	7
Magnesium sulfate (Epsom salts)	Magnesium Sulfur	43	1½	4

suggested amount to the nutrient solution after they are ready for use. Make this stock solution by dissolving three level teaspoons of powdered boric acid and one level teaspoon of manganese chloride or manganese sulfate in 1½ gallons of water. Add ½ pint of this stock solution to 25 gallons of the nutrient solution. There will probably be enough impurities in the various salts and water to provide the needed zinc and copper.

The iron addition should be prepared using 1½ ounces of iron chelate in 1 gallon of water. Use ½ pint of this stock solution to 25 gallons of nutrient solution. Addition of iron must be made once or twice a week. Alternate sources of iron are iron tartrate, iron citrate or ferrous sulfate. Use four level teaspoons of one of these sources per gallon of water and use as directed for the chelated iron.

Remember to keep the stock solutions out of the light, either by using opaque containers or storing them in dark places. This will prevent troublesome algae growth.

LARGE SCALE HYDROPONICS

For commercial type enterprises, hydroponics becomes more complex and requires close detailed attention. It is not an inexpensive way to grow plants, although this will vary with the sophistication of the equipment, the size of the operation and the experience of the operator.

Commercial enterprises commonly prefer the "closed" system of hydroponics. The nutrient solution is periodically pumped from holding containers up through pea gravel beds in which the plants are rooted. Gravity drainage returns the solution to the tank, where it remains until the next pumping cycle. The frequency of pumping is related to the size of the plants, temperature, relative humidity and the drainage rate.

Because of the large investment in commercial hydroponic systems, it is wise to monitor the nutrient concentration and supplement it as needed. Usually the solution is discarded after two or three weeks and completely replaced. Environmental regulations dictate the proper disposal procedure for this solution, and local regulations should be adhered to in all cases. Extended use of the solution can be accomplished if a careful monitoring system is established and proper nutrient salt additions are made.

Since hydroponic solutions are usually common to all of the plants in the system, it is a possible means of rapidly spreading disease, once an infestation occurs. Also, if the solution is not changed regularly, there is a good chance of specific salt buildup. This may come from the pipelines, the water source or from impurities in the chemicals used. These are some of the possible problems associated with hydroponic culture.

It is possible to buy commercial preparations of premixed salts for hydroponics. No single formula has been determined to best fit all plants and all growing conditions. If the operator desires to extend the use of the solutions, then he will need a supply of individual salts to add when the laboratory analysis of the solution indicates this is necessary.

FORMULA FOR LARGE SCALE SYSTEM

The formula given in this chapter has been demonstrated as a successful one for tomato production. It is given here as a basic guide, but many successful operations use an entirely different formula.

Table 11-2. A Commonly Used Stock Solution (Stock Solution 1)

Salt	For 50 Gallons	For 10 Liters
	(lbs.)	(gm)
Potassium nitrate (KNO_3)	21	503
Potassium phosphate (KH_2PO_4)	12	288
Magnesium sulfate ($MgSO_4 \cdot 7H_2O$)	21	503
Micronutrient concentrate	5 gal.	1,000 ml

Two liquid stock solutions are prepared, since certain salts will precipitate when mixed together in concentrated form. These stock solutions can be stored for moderate periods of time.

Add salts, fill container with water and mix thorougly to dissolve. Add micronutrient concentrate (see Table 11-3) with last portion of water.

Table 11-3. A Commonly Used Micronutrient Concentrate

Salt	For 50 Gallons	For 10 Liters
	(gm)	(gm)
Boric acid (H_3BO_3)	54	2.8
Manganese sulfate ($MnSO_4 \cdot H_2O$)	28	1.5
Zinc sulfate ($ZnSO_4 \cdot 7H_2O$)	4	0.2
Copper sulfate ($CuSO_4 \cdot 5H_2O$)	1	0.05
Molybdic acid ($H_2MoO_4 \cdot H_2O$)	0.5	0.03

Add boric acid to $\frac{1}{3}$ volume of water, boil until dissolved and then cool. Next, dissolve other salts in container with about $\frac{2}{3}$ volume of water. After the salts are dissolved, add boric acid solution and bring to final volume.

Table 11-4. A Commonly Used Stock Solution Containing Nitrogen, Calcium and Iron (Stock Solution 2)

Salt	For 50 Gallons	For 10 Liters
	(lbs.)	(gm)
Calcium nitrate *	45	1079
Sequestrene 330 Fe †	2	48

* Commercial grade fertilizer.
† A Ciba-Geigy Chemical Company product. Other iron chelates may work equally well but should be tested before they are used.

Mix the iron chelate (Sequestrene 330 Fe) thoroughly in a little water and then add to the dissolved calcium nitrate.

The nutrient solution is prepared from the two stock solutions. The final dilution is one part of each stock solution to 200 parts water. Table 11-5 shows the amount of stock solution to use for three final nutrient solution volumes:

Table 11-5. The Amount of Stock Solutions Needed for Different Volumes

Stock Solution	Final Volume		
	20 Liters	100 Gallons	1,000 Gallons
Solution 1	100 ml	2 qt.	5 gal.
Solution 2	100 ml	2 qt.	5 gal.

Do not mix the concentrated solutions together without diluting them since calcium phosphate will precipitate.

The final concentration of nutrient elements in the final solution, based upon the 200 to 1 dilution, is as follows:

	N	P	K	Ca	Mg	S	Fe	B	Mn	Zn	Cu	Mo
ppm	119	30	140	100	24	32	2.5	0.25	0.25	0.025	0.01	0.005
me/1	8.5	1	3.5	5	2	2						

It is helpful to know this concentration since laboratories often report concentrations in parts per million or milliequivalents per liter. This will also provide a base from which to calculate how much nutrient needs to be added for replenishment of nutrient in the solution or for a salt substitution. For example, if the water used in the system contains 100 ppm calcium, then calcium nitrate can be left out and a substitute salt such as ammonium nitrate used. Since calcium nitrate supplies 5 me of calcium and nitrogen, then 5 me of nitrogen is needed to supply that which would come from calcium nitrate. Table 11-6 shows the amount of salt required to add to 50 gallons of stock solution to give 1 me/liter when diluted 200-fold. Note this requires 3.34 pounds of ammonium nitrate per me. Thus for 5 me, it will take 16.7 lbs. of ammonium nitrate in stock solution 2 to substitute for the calcium nitrate.

Some basic guidelines must be observed in hydroponic culture. The items covered in this chapter are not intended to be a complete list or to include all crops grown and systems used. Observance of the guidelines listed, however, will serve as a help in avoiding obvious problems.

MONITORING NUTRIENT SOLUTIONS

The nutrient formulas can be quite well monitored by the use of a salt bridge or conductivity meter. Since this instrument measures

Table 11-6. Suggested Salts for Formula Modifications and Amounts to Add

Salt	Amount Required to Add to 50 Gallons Stock Solution to Give 1 me/liter When Diluted 200-fold	
	(nutrient)	(lbs.)
Potassium phosphate (KH_2PO_4)	P	11.4
	K	11.4
Diammonium phosphate [$(NH_4)_2HPO_4$]	P	11.0
	N	5.5
Phosphoric acid (H_3PO_4—52% P_2O_5)	P	11.4
Phosphoric acid (85% H_3PO_4—reagent grade)	P	9.6
Calcium nitrate (fertilizer grade)	Ca	9.0
	N	7.5
Calcium nitrate [$Ca(NO_3)_2 \cdot 4H_2O$—reagent grade]	Ca	9.8
	N	9.8
Ammonium nitrate (NH_4NO_3)	N	3.3
Potassium nitrate (KNO_3)	N	8.4
	K	8.4
Nitric acid (70% HNO_3)	N	7.5
Urea [$CO(NH_2)_2$]	N	2.5

the electrical conductivity of the solution which, in turn, is a function of the nutrients in solution, it provides a guide as to the decline in mineral nutrients as they are used by the plants. A base point should be established by measuring the solution conductivity immediately after the salts are added and mixed thoroughly. Additional measurements should be made every day or, depending upon the size of the plants, every two or three days.

The electrical conductivity (EC) measures the concentration of the total ions in solution, not individual ions. Experience has shown, however, that there is a good correlation between EC and nitrate-nitrogen, phosphate-phosphorus and potassium. As the EC decreases, fertilizer salt or stock solution should be added to bring the EC back to the base level. Bring the volume of solution nearly to the original level with water before making the EC reading, particularly with larger, faster-growing plants.

The quantity of stock solution to add is related to the growth rate of the plants. Usually about 1 gallon of stock solution is needed per day per 1,000 gallons of solution volume for 6 to 8 foot tall plants. During the first month, while plants are small, the requirements are $\frac{1}{4}$ or less. Adjustments are made based upon the EC reading and experience. Make sure that the solution is thoroughly mixed before EC measurements are made.

Water quality is an important factor to consider in hydroponic culture. Where the water may have a high content of salt, it may limit the time it can be used, if at all, and it may contribute some salts in excess or of a toxic nature. A good practice to follow to avoid water problems is to have a water quality test made by a reliable laboratory.

Solution volume should be kept to 4 or 5 gallons per plant for tomatoes. This may be somewhat less for other plants, but care should be taken to provide adequate solution. Quick changes in solution concentration are thus avoided, and the margin of safety in culture is extended.

Another monitoring system that allows for the evaluation of the tissue concentration of nutrients should be periodically performed. Again, a reliable laboratory should be used and general guidelines for sampling are described in Chapter 9. Critical levels are listed for various plants.

Consultation with the proper authorities at the Cooperative Extension or University will be of assistance. Some references concerning this method of growing plants are included in the supplementary reading suggestions, which may also be a valuable aid.

SUPPLEMENTARY READING

1. "Growing Plants in Nutrient Cultures." John G. Seeley. *Horticulture Science.* August 1974.
2. *Growing Plants in Solution Culture.* E. Epstein and B. A. Krantz. AXT-196. University of California. 1972.
3. *Hydroponics: A Guide to Soilless Culture Systems.* Hunter Johnson, Jr. Leaflet 2947. University of California. 1977.
4. *Soilless Growth of Plants,* Second Edition. Carleton Ellis and M. W. Swaney. Van Nostrand Reinhold Company. 1974.

Chapter 12

BENEFITS OF FERTILIZERS TO
THE ENVIRONMENT

Research shows that the judicious use of fertilizer can improve man's environment, while continued heavy use in excess of crop requirements can lead to potential environmental problems. This chapter will discuss briefly how the judicious use of fertilizer leads to an improvement of man's environment and how excessive use can cause environmental problems. No attempt will be made to completely cover all of the positive and negative influences that fertilizers can exert on the environment. Also discussed are practices which can be used to help match fertilizer inputs to crop requirements to achieve high yield agriculture with protection for the environment.

Recent years have seen the creation of numerous agencies and bureaus whose roles directly or indirectly have a bearing on fertilizers, their manufacture, storage, handling and transportation. The authority of these organizations is subject to change. Since this is the case, no attempt will be made to give detailed information on the scope of the functions of these agencies and bureaus. Names and addresses are given, and any company or individual working closely in the areas affected by these organizations should write for current detailed information.

HOW THE JUDICIOUS USE OF FERTILIZER
IMPROVES THE ENVIRONMENT

Judicious use of fertilizer improves the environment in several ways. It:

1. Gives people cleaner air.
2. Cuts down on soil erosion, which leads to cleaner surface waters.
3. Reduces pollution of ground waters.
4. Leaves more land for open spaces and recreational purposes.
5. Provides a means for disposing of degradable wastes.

Fig. 12-1. Heavy top growth of well-fed crops reduces the pounding effect of falling water.

Gives People Cleaner Air

Growing plants, through a process called photosynthesis, convert carbon dioxide into carbohydrates, and at the same time give off oxygen. These plants take in the carbon dioxide through microscopic openings in their leaves, called stomata, and give off life-supporting oxygen. Man, animals and industry consume large quantities of oxygen and generate large quantities of carbon dioxide. The plants on the earth and in the oceans maintain the oxygen:carbon dioxide ratio in the air. Air contains about 20 percent oxygen and .03 percent carbon dioxide. This balance is a delicate one. Since larger crops use more carbon dioxide and give off more oxygen, a well-fertilized crop is more productive of oxygen than a poorly fertilized crop. Judicious fertilization stimulates crop growth and thus indirectly increases the amount of oxygen given off by the crop.

The 200 million people in the United States require an estimated 600 million pounds of oxygen a day. We use this vital element in varying amounts depending upon our activities. A person working in an office all day uses about 2 pounds of oxygen, whereas a person doing physical labor may use 6 pounds or more per day.

A well-fertilized field of grain will give off 4 to 5 tons of oxygen per acre annually. It also takes in some 6 tons of carbon dioxide. A healthy, well nourished orange grove uses up 6 or more tons of carbon dioxide per acre while generating about 4 tons of oxygen—enough oxygen for nine people for a year.

Still more important to man, however, is the fact that productive soils remove carbon monoxide from the air. Scientists have identified the *Aspergillus* and *Penicillium* types of fungi as responsible for carbon monoxide removal. These organisms, most abundant in fertile soils, are helping to provide cleaner air by using for their metabolism much of the 200 million tons of carbon monoxide created annually by man. This does not mean that carbon monoxide is any less toxic to man. It does mean that in spite of man's activities, carbon monoxide will not build up in the atmosphere to a dangerous level except perhaps on a localized basis.

Gas chromatographic studies also show that fertile soils have the capacity to absorb substantial quantities of sulfur dioxide and hydrogen sulfide. These experiments involving both steam sterilized and non-steam sterilized soils indicate that soil microorganisms play little, if any, part in the absorption of sulfur dioxide and hydrogen sulfide. The research reveals that moist, fertile soils can absorb from 9.3 to 66.2 pounds of sulfur dioxide per acre. In brief, these investigations indicate that soil is an important natural sink for gaseous atmospheric pollutants and may prove valuable for purification of industrial emissions heavily polluted by sulfur gases.

Cuts Down on Soil Erosion

Well-fertilized crops have both extensive tops and roots. A well-developed and extensive top growth reduces the pounding effect of natural raindrops or sprinklers. The energy of the drops is dissi-

Fig. 12-2. Soil erosion can be serious when the soil is not held in place by roots of well-fertilized crops.

pated so that instead of breaking down the structure of the soil surface the water trickles down into the soil. Runoff is reduced and erosion is minimized. Well-fertilized crops with extensive root systems also hold the soil in place against water runoff. With minimal soil erosion, streams run clear and clean.

Research has shown that soil erosion is the major way phosphates are lost from the soil. The phosphorus is carried off as adsorbed and precipitated phosphate. Soil sediment also contains nitrogen and other nutrients so any reduction in erosion also reduces the levels of these nutrients in surface water.

Reduces Pollution of Ground Waters

Nitrates, along with other soluble nutrients such as chlorides and sulfates, are associated with underground water pollution. As previously mentioned, phosphates can normally be eliminated from discussion since even completely water-soluble forms, when applied to the soil, are rapidly converted to water-insoluble forms and therefore do not move with the soil solution. Phosphate stays essentially where it is placed in the soil, and there is little or no leaching. This is why fertilizer-phosphate placement within the root zone is a commonly recommended practice. If erosion is controlled, phosphates stay where placed.

Regardless of the form of nitrogen applied to the soil, essentially all of it will eventually be converted to the nitrate form. The nitrate form moves downward, laterally or upward, depending on the movement of soil moisture. If the nitrates in the soil solution move below the root zone, they may eventually end up in the ground water. There are natural processes such as denitrification that may reduce or eliminate the contamination.

Crops with extensive root systems—the kind that develop when well fertilized—utilize the soluble nutrients including nitrates as the soil solution moves either downward or upward within the root zone of the plants. The deeper and more extensive the root system, the more effective the crop is in keeping soluble nutrients from reaching the ground waters.

An active and strong root system acts like a wick. It not only takes up water in the immediate area, but also, through upward capillary action, draws water along with soluble materials from deep in the soil. This action helps reduce the potential for pollution of ground waters.

Leaves More Land for Open Spaces and Recreational Purposes

Fertilizer is responsible for approximately 35 percent of agriculture production. Without the use of fertilizer, farmers would have to farm many more acres of land to produce food and fiber, and this would mean less land for open spaces and for recreational purposes. More cultivated land would also mean more soil erosion.

Several studies in the United States have been conducted to determine the degree fertilizer substitutes for land. Donald Ibach of the USDA estimated that 1 ton of nutrients applied annually would substitute for 9.4 acres of land. The degree of substitution varies, of course, depending on the rate of fertilizer application, yield response, types of crop management practices and other variables. Table 12-1, based on USDA statistics, shows how total farm output and production per acre are related to fertilizer usage.

Recreational areas are very important to much of the population which is crowded into large urban centers. Parks, playgrounds, golf

Table 12-1. Changes in Fertilizer Usage, Farm Output, Production per Acre, Crop Land Use and Man-Hours in U.S. Agriculture*

Year	Fertilizer Nutrients Used	Total Farm Output	Crop Production per Acre	Crop Land per Crop	Man-Hours of Farm Work
1950	100	100	100	100	100
1955	151	112	107	101	85
1960	184	123	128	95	65
1965	258	133	145	89	51
1970	396	140	148	89	43
1975	434	155	162	97	35

*Relative percentages based upon 1950 figures.

courses, street plantings and home gardens are greatly appreciated by these people.

Good turf is a basic ingredient for recreational and esthetic needs. Well-grown turf also protects environmental quality. A good grass cover prevents dust from blowing, stops water erosion, filters sediment from storm runoff and keeps streets cleaner. Grass also causes considerable evaporative cooling, which in hot areas may reduce home air conditioning requirements during hot nights.

Trees and shrubs can provide much needed shade, can reduce noise pollution and can provide visual relief among city buildings. These plantings, like grass, depend on good growth for their value. Such growth requires careful maintenance, including the proper use of fertilizer.

More land is available for wildlife, due to intensive farming with fertilizers which has reduced the need for land. Some land is available for elk refuges, wildfowl sanctuaries, wilderness areas, parks and forest preserves. There are also lakes for fishing. Since farmers do not have to farm every acre, they can leave the rougher land for wildlife. As crop acres are fertilized and more residues are produced, there is more feed for this wildlife.

Forest fertilization is a standard practice in many areas of the U.S. This helps grow more timber for wood and paper products to meet expanding demands. Trees are a renewable resource. Structural wood does not consume limited resources such as metals. Paper is biodegradable and easily recycled. Producing tree crops quicker and at a lower cost by proper fertilization is a "must" for the protection of

Fig. 12-3. Through the judicious use of fertilizer, more land can be used for recreation areas and wildlife preserves.

the environment. The more trees are farmed like the food crops, the more land will be available for national forest preserves.

Man is the most important animal in his environment. Protecting the quality of the soil, water and air provides more opportunity for each person to enjoy more quality of life. Food, water and air are man's most basic needs.

To have ample food means to have it available at a price one can afford. Individuals can no longer grow their own food except where a backyard is available for gardening. Food at a price one can afford means food produced as cheaply as possible. Thus, inexpensive food has always been a national concern. Such a food supply is made available by a strong agricultural industry working as efficiently as possible.

Fertilizer is the most profitable input of all those which farmers use in crop production. Thus, fertilizer is basic to having enough quality food for everyone at a price everyone can afford. It lowers the unit cost of producing a crop which lowers the cost of food to the consumer. With this basic need met, man can look forward to a better quality of life.

Provides a Means for Disposing of Degradable Wastes

Despite society's concern about environmental problems, only recently has the tremendous capacity of the soils to degrade waste been exploited. The soil is a kind of massive machine for keeping the chemical material of the planet in circulation. Its framework is a collection of individual soil particles, each tiny particle possessing a remarkably large surface. An ounce of soil may have surfaces totalling 250,000 square feet. The spaces between the soil particles harbor an almost incalculable population of microbial life. A gram of productive soil from the temperate regions may have 4 billion bacteria, 20 million actinomycetes, $\frac{1}{4}$ million protozoa and $1\frac{1}{2}$ million algae and fungi. The microbial activity in one acre of soil expends about the same amount of energy as 10,000 people.

When it's all said and done, soil still remains the best and about the only way of disposing of the masses of degradable waste that man generates annually.

MATCHING FERTILIZER INPUTS TO CROP NEEDS

Judicious fertilization means matching fertilizer inputs to crop and soil needs. Judicious fertilization calls for:
1. Using the RIGHT NUTRIENTS.
2. Using the RIGHT AMOUNT of the nutrients.
3. Applying the nutrients in the RIGHT PLACE.
4. Applying the nutrients at the RIGHT TIME for the crops.

There are several "tools" available to help the grower match fertilizer inputs to crop and soil needs. These include soil and tissue testing discussed in Chapter 9 and plant food requirements of various crops discussed in Chapter 4. Past yield production records, as well as managerial abilities of a grower, are important considerations to be used in arriving at which fertilizer and how much of it to use to achieve maximum production without using excessive quantities.

AGENCIES AND BUREAUS VESTED WITH FERTILIZER CONTROLS AND REGULATIONS

The scope and activities of fertilizer control with respect to meeting

Fig. 12-4.   Fertilizers are credited with over a third of our food production.

fertilizer guarantees is covered in Chapter 13. As previously mentioned, recent years have seen the creation of several new agencies, bureaus and organizations, most of which deal in one way or another with the quality of our environment. These organizations can have an effect on the manufacture, storage, handling, transportation and use of fertilizer. Listed below are their names and addresses. Current detailed information can be obtained by writing directly to these organizations:

National Office
Occupational Safety and Health Administration (OSHA)
U.S. Department of Labor
Washington, DC 20210

Regional Office
Occupational Safety and Health Administration (OSHA)
10353 Federal Building
P.O. Box 36017
San Francisco, CA 94101

Environmental Protection Agency (EPA)
Regional Office No. 9
100 California Street
San Francisco, CA 94111

Department of Transportation (DOT)
Office of Hazardous Materials
2100 Second Street, S.W.
Washington, DC 20590

Chemical Transportation Center (CHEMTREC)
1825 Constitution Avenue, N.W.
Washington, DC 20009

California State Water Resources Control Board
Division of Planning and Research
1416 Ninth Street
Sacramento, CA 95814

California State Solid Waste Management Board
State Department of Public Health
Resources Building
1416 Ninth Street, Room 1335
Sacramento, CA 95814

State Air Resources Board
1709 Eleventh Street
Sacramento, CA 95814

Division of Industrial Safety
Department of Industrial Relations
455 Golden Gate Avenue
San Francisco, CA 94101

SUPPLEMENTARY READING

1. *Contributions of Fertilizer to the Economy and the Environment.* L. B. Nelson. TVA Publication. 1972.
2. *Facts from Our Environment.* Potash Institute of North America. Special Bulletin. 1972.
3. "Nitrate Breakdown." *Agricultural Research.* May 1970.
4. "Nitrogen Facts and Fallacies." W. L. Garman. *Plant Food Review,* No. 1. 1969.
5. "Sorption of Gaseous Atmospheric Pollutants by Soils." *Soil Sci.* Vol. 116, No. 4: 313-319. 1973.

Chapter 13

WESTERN LAWS RELATING TO
FERTILIZING MATERIALS

The Association of American Plant Food Control Officials was formed in 1946. The objectives of the Association are to promote uniform and effective legislation, definitions, rulings and enforcement of laws relating to the control of sale and distribution of mixed fertilizers on the continent of North America. These officials function to protect the fertilizer industry and the farmer alike against any who seek to practice unfair dealings.

The membership of the association consists of:

1. Officers charged by law with the active execution of the laws regulating the sale of commercial fertilizer and fertilizer materials.
2. Deputies who may be designated by the officials named in Section 1.
3. Research workers employed by State, Dominion or Federal agencies who are engaged in the investigation of fertilizers.

This organization studies through committees or investigation any problems which may arise concerning the various points listed as objectives. Regular meetings are held at least once a year. Through their Association, the Fertilizer Control Officers have been able to work for improved laws and definitions and to discuss mutual problems of enforcement.

During recent years a proposed "Model State Fertilizer Bill" was drafted in an effort to bring about uniformity in fertilizer regulation. The provisions of the fertilizer laws in most states and provinces follow this "Model Bill." However, regulations covering registration, labeling, reporting of tonnage, violations and tonnage fees vary considerably in the different areas. It is recommended that those desiring copies of laws which apply to the particular area in which they operate write to the governing Fertilizer Control Official. Addresses are given below.

For information about states not listed in this chapter, write to the Association of American Plant Food Control Officials, Department of Biochemistry, Purdue University, West Lafayette, Indiana 47907. The association publishes annually an official publication which contains the constitution and by-laws, fertilizer terms, annual report and a complete list of state fertilizer control officials.

Alaska
Division of Agriculture
Box 1088
Palmer, AK   99645

Arizona
Office of State Chemist
Agr. Experiment Station
P.O. Box 1586
Mesa, AZ   85201

California
Agricultural Chemistry and Feed
1220 "N" Street
Sacramento, CA   95814

Canada
Chief, Feed and Fertilizer Section
Sir John Carling Building
Canada Department of Agriculture
Ottawa, Ontario
Canada

Colorado
Department of Agriculture
2331 West 31st Avenue
Denver, CO   80211

Hawaii
Chief, Commodities Branch
Department of Agriculture
P.O. Box 5425
Honolulu, HI   96814

Idaho
Director, Bureau of Plant Ind.
Department of Agriculture
P.O. Box 790
Boise, ID   83701

Montana
Montana Department of
Agriculture
McCall Hall
Montana State University
Bozeman, MT   59601

Nevada
Chief Chemist
Department of Agriculture
P.O. Box 11100
Reno, NV   89510

New Mexico
Chief, Division
Feed, Seed & Fertilizer
P.O. Box 3150
Las Cruces, NM   88003

Oregon
Administrator, Plant Division
Department of Agriculture
635 Capitol Street, N.E.
Salem, OR   97310

Utah
State Chemist
Department of Agriculture
Room 412
State Capitol Building
Salt Lake City, UT   89114

Washington
State Chemist
Grain and Chemical Division
Department of Agriculture
P.O. Box 2113
Yakima, WA   98902

Wyoming
Director, Division of Plant Industry
2219 Carey Avenue
Cheyenne, WY   82002

UNIFORM STATE FERTILIZER BILL*

NOTE—Although this Bill and Regulations have not been passed into

---

* The model law reproduced by courtesy of AAPFCO.

law in all states, the subject matter covered herein does represent the official policy of this Association.

AN ACT to regulate the sale and distribution of commercial fertilizers in the State of .... BE IT ENACTED BY the Legislature of the State of ....

Section 1. Title.

This Act shall be known as the "... Fertilizer Law of 19.................."

Section 2. Enforcing Official.

This Act shall be administered by the .... of the State of ...., hereinafter referred to as the "..............................."

Section 3. Definitions of Words and Terms.

When used in this Act:

A. The term "commercial fertilizer" means any substance containing one or more recognized plant nutrient(s) which is used for its plant nutrient content and which is designed for use or claimed to have value in promoting plant growth, *except* unmanipulated animal and vegetable manures, marl, lime, limestone, wood ashes and gypsum, and other products exempted by regulation of the ....

    (1) A "fertilizer material" is a commercial fertilizer which either:
        a. Contains important quantities of no more than one of the primary plant nutrients (nitrogen, phosphoric acid and potash), or
        b. Has approximately 85% of its plant nutrient content present in the form of a single chemical compound, or
        c. Is derived from a plant or animal residue or by-product or a natural material deposit which has been processed in such a way that its content of primary plant nutrients has not been materially changed except by purification and concentration.
    (2) A "mixed fertilizer" is a commercial fertilizer containing any combination or mixture of fertilizer materials.
    (3) A "specialty fertilizer" is a commercial fertilizer distributed primarily for nonfarm use, such as home gardens, lawns, shrubbery, flowers, golf courses, municipal parks, cemeteries, greenhouses and nurseries.
    (4) A "bulk fertilizer" is a commercial fertilizer distributed in a non-packaged form.

B. The term "brand" means a term, design or trade mark used in connection with one or several grades of commercial fertilizer.

C. Guaranteed Analysis:
    (1) Until the .... prescribes the alternative form of "guaranteed analysis" in accordance with the provisions of subparagraph (2) hereof, the term "guaranteed

analysis" shall mean the minimum percentage of plant nutrients claimed in the following order and form:

a. Total Nitrogen (N) .... percent
   Available Phosphoric Acid (P_2O_5) .... percent
   Soluble Potash (K_2O) .... percent

b. For unacidulated mineral phosphatic materials and basic slag, bone, tankage and other organic phosphate materials, the total phosphoric acid and/or degree of fineness may also be guaranteed.

c. Guarantees for plant nutrients other than nitrogen, phosphorus and potassium may be permitted or required by regulation of the .... The guarantees for such other nutrients shall be expressed in the form of the element. The sources of such other nutrients (oxides, salt, chelates, etc.) may be required to be stated on the application for registration and may be included as a parenthetical statement on the label. Other benethetical substances or compounds, determinable by laboratory methods, also may be guaranteed by permission of the .... and with the advice of the Director of the Agricultural Experiment Station. When any plant nutrients or other substances or compounds are guaranteed, they shall be subject to inspection and analysis in accord with the methods and regulations prescribed by ....
....

d. Potential basicity or acidity expressed in terms of calcium carbonate equivalent in multiples of one hundred pounds per ton, when required by regulation.

(2) When the .... finds, after public hearing following due notice, that the requirement for expressing the guaranteed analysis of phosphorus and potassium in elemental form would not impose an economic hardship on distributors and users of fertilizer by reason of conflicting labeling requirements among the states, he may require by regulation thereafter that the "guaranteed analysis" shall be in the following form:

Total Nitrogen (N) .... _____ percent
Available Phosphorus (P) .... _____ percent
Soluble Potassium (K) .... _____ percent

Provided, however, that the effective date of said regulation shall be not less than six months following the issuance thereof, and provided, further, that for a period two years following the effective date of said regulation the equivalent of phosphorus and potassium may also be shown in the form of phosphoric acid and potash; provided, however, that after the effective date of a regulation issued under the provisions of this section, requiring that phosphorus and potassium be

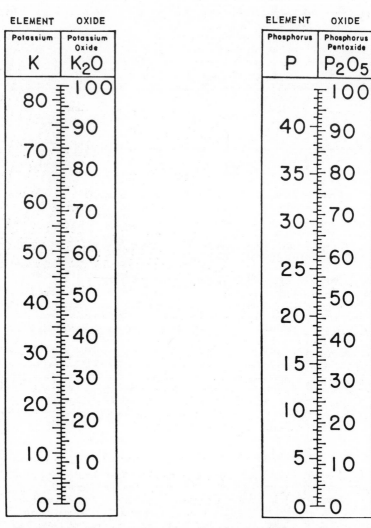

Fig. 13-1. Fertilizer conversion chart for changing oxide to element and vice versa (pounds or percent).

shown in the elemental form, the guaranteed analysis for nitrogen, phosphorus and potassium shall constitute the grade.

D. The term "grade" means the percentage of total nitrogen, available phosphorus or phosphoric acid, and soluble potassium or soluble potash stated in whole numbers in the same terms, order and percentages as in the guaranteed analysis. Provided however that fertilizer materials, bone meal, manures and similar raw materials may be guaranteed in fractional units.

E. The term "official sample" means any sample of commercial fertilizer taken by the .... or his agent and designated as "official" by the ....

F. The term "ton" means a net weight of two thousand pounds avoirdupois.

G. The term "percent" or "percentage" means the percentage by weight.

H. The term "person" includes individual, partnership, association, firm and corporation.

I. The term "distributor" means any person who imports, consigns, manufactures, produces, compounds, mixes or blends commercial fertilizer, or who offers for sale, sells, barters or otherwise supplies commercial fertilizer in this state.

J. The term "registrant" means the person who registers commercial fertilizer under the provisions of this Act.

K. The term "licensee" means the person who receives a license to distribute a commercial fertilizer under the provisions of this Act.

L. The term "label" means the display of all written, printed or graphic matter upon the immediate container or statement accompanying a commercial fertilizer.

M. The term "labeling" means all written, printed or graphic matter, upon or accompanying any commercial fertilizer, or advertisements, brochures, posters, television and radio announcements used in promoting the sale of such commercial fertilizers.

N. The term "investigational allowance" means an allowance for variations inherent in the taking, preparation and analysis of an official sample of commercial fertilizer.

Section 4. Registration (Option A).

A. Each brand and grade of commercial fertilizer shall be registered before being distributed in this State. The application for registration shall be submitted to the .... on a form furnished by the ...., and shall be accompanied by a fee of $.................... per brand and grade except those fertilizers sold in packages of 10 pounds or less shall be registered at a fee of twenty-five dollars each. Upon approval by the .... a copy of the registration shall be

furnished to the applicant. All registrations expire on ....
.... of the following year.
The application shall include the following information:
(1)   The net weight;
(2)   The brand and grade;
(3)   The guaranteed analysis;
(4)   The name and address of the registrant.

B.   A distributor shall not be required to register any commercial fertilizer which is already registered under this Act by another person, providing the label does not differ in any respect.

C.   A distributor shall not be required to register each grade of commercial fertilizer formulated according to specifications which are furnished by a consumer prior to mixing, but shall be required to register his firm in a manner and at a fee as prescribed in the regulations by the ...., and to label such fertilizer as provided in Section 5(b).

Licensing (Option B)

A.   No person whose name appears upon the label of a commercial fertilizer shall distribute that fertilizer, except specialty fertilizers, to a non-licensee until a license to distribute has been obtained by that person from the .... upon payment of a $.................... fee. All licenses expire on the thirtieth day of June of each year.

B.   An application for license shall include:
(1)   The name and address of licensee;
(2)   The name and address of each bulk distribution point in the state not licensed for fertilizer manufacture and distribution.
The name and address shown on the license shall be shown on all labels, pertinent invoices and bulk storage for fertilizers distributed by the licensee in this state.

C.   The licensee shall inform the .... in writing of additional distribution points established during the period of the license.

D.   No person shall distribute in this state a specialty fertilizer until it is registered by the manufacturer or distributor with the .... An application in duplicate for each brand and product name of each grade of specialty fertilizer shall be made on a form furnished by the .... and shall be accompanied with a fee of $.................... for each brand and product name of each grade. Labels for each brand and product name of each grade shall accompany the application. Upon the approval of an application by the ...., a copy of the registration shall be furnished the applicant. All registrations expire on the thirtieth day of June of each year.

E.   An application for registration shall include the following:
(1)   Name and address of the manufacturer or distributor;

(2)   The brand and product name;
(3)   The grade;
(4)   The guaranteed analysis;
(5)   The package sizes for persons who package fertilizers only in containers of ten pounds or less.

Section 5. Labels.

A.   Any commercial fertilizer distributed in this State in containers shall have placed on or affixed to the container a label setting forth in clearly legible and conspicuous form the information required by Section 4 (A) (1), (2), (3) and (4) of this Act. In case of bulk shipments, this information in written or printed form shall accompany delivery and be supplied to the purchaser at time of delivery.

B.   A commercial fertilizer formulated according to specifications which are furnished by a consumer prior to mixing shall be labeled to show the net weight, guaranteed analysis, and the name and address of the distributor.

Section 6. Inspection Fees.

A.   There shall be paid to the .... for all commercial fertilizers distributed in this State an inspection fee at the rate of .... cents per ton: Provided, that sales to manufacturers or exchanges between them are hereby exempted. Fees so collected shall be used for the payment of the costs of inspection, sampling and analysis, and other expenses necessary for the administration of this Act.
    On individual packages of commercial fertilizer containing 10 pounds or less, there shall be paid in lieu of the annual registration fee of $..................... per brand and grade and the .... cents per ton inspection fee, an annual registration fee and inspection fee of twenty-five dollars for each brand and grade sold or distributed. Where a person sells commercial fertilizer in packages of 10 pounds or less and in packages over 10 pounds, this annual registration and inspection fee of twenty-five dollars shall apply only to that portion sold in packages of 10 pounds or less, and that portion sold in packages over 10 pounds shall be subject to the same inspection fee of .... cents per ton as provided in this Act.

B.   Every person who distributes a commercial fertilizer in this State shall file with the .... on forms furnished by .... a quarterly statement for the periods ending September 30, December 31, March 31 and June 30, setting forth the number of net tons of each commercial fertilizer distributed in this state during such quarter. The report shall be due on or before the thirtieth day of the month following the close of each quarter and upon such statement shall pay the inspection fee at the rate stated in paragraph (a) of this section.

If the tonnage report is not filed and the payment of inspection fee is not made within 30 days after the end of the quarter, a collection fee amounting to 10 percent (minimum $10.00) of the amount shall be assessed against the registrant, and the amount of fees due shall constitute a debt and become the basis of a judgment against the registrant.

C. When more than one person is involved in the distribution of a commercial fertilizer, the last person who has the fertilizer registered and who distributes to a nonregistrant (dealer or consumer) is responsible for reporting the tonnage and paying the inspection fee, unless the report and payment is made by a prior distributor of a fertilizer.

Section 7. Tonnage Reports.

A. The person transacting, distributing or selling commercial fertilizer to a nonregistrant shall mail the .... a report showing the county of the consignee, the amounts (tons) of each grade of commercial fertilizer, and the form in which the fertilizer was distributed (bags, bulk, liquid, etc.). This information shall be reported by *one* of the following methods: (1) submitting a summary report approved by the .... on or before the .... day of each month covering shipments made during the preceding month; or (2) submitting a copy of the invoice within .... business days after shipment. No information furnished the .... under this section shall be disclosed in such a way as to divulge the operation of any person.

Section 8. Inspection, Sampling, Analysis.

A. It shall be the duty of the ...., who may act through his authorized agent, to sample, inspect, make analyses of, and test commercial fertilizers distributed within this State at any time and place and to such an extent he may deem necessary to determine whether such commercial fertilizers are in compliance with the provisions of this Act. The ...., individually or through his agent, is authorized to enter upon any public or private premises or carriers during regular business hours in order to have access to commercial fertilizers subject to the provisions of this Act and the rules and regulations pertaining thereto, and to the records relating to their distribution.

B. The methods of analysis and sampling shall be those adopted by the .... from sources such as the AOAC Journal.

C. The ...., in determining for administrative purposes whether any commercial fertilizer is deficient in plant food, shall be guided solely by the official sample as defined in paragraph (E) of Section 3, and obtained and analyzed as provided for in paragraph (B) of this section.

D. The results of official analysis of commercial fertilizers and portions of official samples shall be distributed by the .... as provided in the regulations.

Section 9. Plant Food Deficiency.

A. Penalty for Nitrogen, Available Phosphoric Acid or Phosphorus and Potash or Potassium—If the analysis shall show that a commercial fertilizer is deficient (1) in one or more of its guaranteed primary plant foods (NPK) beyond the "investigational allowances" as established by regulation, or (2) if the overall index value of the fertilizer is below the level established by regulations, a penalty of .... times the commercial value of such deficiency or deficiencies shall be assessed. When a commercial fertilizer is subject to a penalty under both (1) and (2) the larger penalty shall apply.

B. Penalty for Other Deficiencies—Deficiencies beyond the investigational allowances as established by regulation in any other constituent(s) covered under Section 3 paragraph (c) B, C and D of this Act, which the registrant is required to or may guarantee, shall be evaluated and penalties prescribed therefore by the ....

C. Nothing contained in this section shall prevent any person from

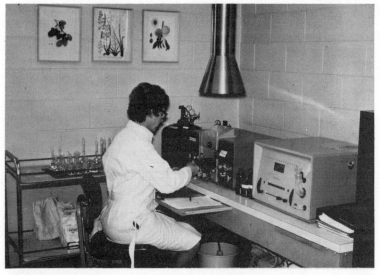

Fig. 13-2. Control of fertilizer quality is maintained through exact laboratory analyses.

appealing to a court of competent jurisdiction praying for judgment as to the justification of such penalties.

D. All penalties assessed under this section shall be paid to the consumer of the lot of commercial fertilizer represented by the sample analyzed within three months after the date of notice from the .... to the registrant, receipts taken therefore and promptly forwarded to the .... If said consumers cannot be found, the amount of the penalty shall be paid to the .... who shall deposit the same (or shall pay said penalty to some local charitable or educational institution).

Section 10. Commercial Value.

For the purpose of determining the commercial value to be applied under the provisions of Section 9 the .... shall determine and publish annually the values per unit of nitrogen, available phosphoric acid and soluble potash in commercial fertilizers in this state. If guarantees are as provided in Section 3 (c) (2) the value shall be per unit of nitrogen, phosphorus, and potassium. The values so determined and published shall be used in determining and assessing penalties.

Section 11. Misbranding.

No person shall distribute misbranded fertilizer. A commercial fertilizer shall be deemed to be misbranded:

A. If its labeling is false or misleading in any particular.

B. If it is distributed under the name of another fertilizer product.

C. If it is not labeled as required in Section 5 of this act and in accordance with regulations prescribed under this act.

D. If it purports to be or is represented as a commercial fertilizer, or is represented as containing a plant nutrient or commercial fertilizer unless such plant nutrient or commercial fertilizer conforms to the definition of identity, if any, prescribed by regulation of the .... ; in the adopting of such regulations the .... shall give due regard to commonly accepted definitions and official fertilizer terms such as those issued by the Association of American Plant Food Control Officials.

Section 12. Adulteration.

No person shall distribute an adulterated fertilizer product. A commercial fertilizer shall be deemed to be adulterated:

A. If it contains any deleterious or harmful ingredient in sufficient amount to render it injurious to beneficial plant life when applied in accordance with directions for use on the label, or if adequate warning statements or directions for use, which may be necessary to protect plant life are not shown upon the label.

    B.   If its composition falls below or differs from that which it is purported to possess by its labeling.

    C.   If it contains unwanted crop seed or weed seed.

Section 13. Publications.

The .... shall publish at least annually and in such forms as he may deem proper: (a) information concerning the distribution of commercial fertilizers, (b) results of analyses based on official samples of commercial fertilizers distributed within the state as compared with the analyses guaranteed under Section 4 and Section 5.

Section 14. Rules and Regulations.

The .... is authorized to prescribe and, after a public hearing following due public notice, to enforce such rules and regulations relating to investigational allowances, definitions, records, and the distribution of commercial fertilizers as may be necessary to carry into effect the full intent and meaning of this Act.

Section 15. Short Weight.

If any commercial fertilizer in the possession of the consumer is found by the .... to be short in weight, the registrant of said commercial fertilizer shall within thirty days after official notice from the .... pay to the consumer a penalty equal to four times the value of the actual shortage.

Section 16. Cancellation of Registrations.

The .... is authorized and empowered to cancel the registration of any brand of commercial fertilizer or to refuse to register any brand of commercial fertilizer as herein provided, upon satisfactory evidence that the registrant has used fraudulent or deceptive practices in the evasions or attempted evasions of the provisions of this Act or any rules and regulations promulgated thereunder: Provided, that no registration shall be revoked or refused until the registrant shall have been given the opportunity to appear for a hearing by the ....

Section 17. "Stop Sale" Orders.

The .... may issue and enforce a written or printed "stop sale, use or removal" order to the owner or custodian of any lot of commercial fertilizer and to hold at a designated place when the .... finds said commercial fertilizer is being offered or exposed for sale in violation of any of the provisions of this Act until the law has been complied with and said commercial fertilizer is released in writing by the ...., or said violation has been otherwise legally disposed of by written authority. The .... shall release the commercial fertilizer so withdrawn when the requirements of the provisions of this Act have been complied with and all costs and expenses incurred in connection with the withdrawal have been paid.

Section 18. Seizure, Condemnation and Sale.

Any lot of commercial fertilizer not in compliance with the provisions of this Act shall be subject to seizure on complaint of the .... to a court of competent jurisdiction in the area in which said commercial fertilizer is located. In the event the court finds the said commercial fertilizer to be in violation of this Act and orders the condemnation of said commercial fertilizer it shall be disposed of in any manner consistent with the quality of the commercial fertilizer and the laws of the State: Provided, that in no instance shall the disposition of said commercial fertilizer be ordered by the court without first giving the claimant an opportunity to apply to the court for release of said commercial fertilizer or for permission to process or relabel said commercial fertilizer to bring it into compliance with this Act.

Section 19. Violations.

A. If it shall appear from the examination of any commercial fertilizer that any of the provisions of this Act or the rules and regulations issued thereunder have been violated, the .... shall cause notice of the violations to be given to the registrant, distributor or possessor from whom said sample was taken; any person so notified shall be given opportunity to be heard under such rules and regulations as may be prescribed by the .... If it appears after such hearing, either in the presence or absence of the person so notified, that any of the provisions of this Act or rules and regulations issued thereunder have been violated, the .... may certify the facts to the proper prosecuting attorney.

B. Any person convicted of violating any provision of this Act or the rules and regulations issued thereunder shall be punished at the discretion of the court.

C. Nothing in this Act shall be construed as requiring the .... or his representative to report for prosecution or for the institution of seizure proceedings as a result of minor violations of the Act when he believes that the public interests will be best served by a suitable notice of warning in writing.

D. It shall be the duty of each .... attorney to whom any violation is reported to cause appropriate proceedings to be instituted and prosecuted in a court of competent jurisdiction without delay.

E. The .... is hereby authorized to apply for and the court to grant a temporary or permanent injunction restraining any person from violating or continuing to violate any of the provisions of this Act or any rule or regulation promulgated under the Act notwithstanding the existence of other remedies at law. Said injunction to be issued without bond.

Section 20. Exchanges Between Manufacturers.

Nothing in this Act shall be construed to restrict or avoid sales or exchange of commercial fertilizers to each other by importers, manufacturers, or manipulators who mix fertilizer materials for sale or as preventing the free and unrestricted shipments of commercial fertilizer to manufacturers or manipulators who have registered their brands as required by the provisions of this Act.

Section 21. Constitutionality.

If any clause, sentence, paragraph or part of this Act shall for any reason be judged invalid by any court of competent jurisdiction, such judgment shall not affect, impair, or invalidate the remainder thereof but shall be confined in its operation to the clause, sentence, paragraph, or part thereof directly involved in the controversy in which such judgment shall have been rendered.

Section 22. Repeal.

All laws and parts of laws in conflict with or inconsistent with the provisions of this Act are hereby repealed.

Section 23. Effective Date.

This Act shall take effect and be in force from and after the first day of ....

RULES AND REGULATIONS

Under the Uniform State Fertilizer Bill by the .... of the State of ....

Pursuant to due publication and notice of opportunity for a public hearing, the .... has adopted the following regulations.

1. Plant Nutrients in Addition to Nitrogen, Phosphorus and Potassium.

Other Plant Nutrients, when mentioned in any form or manner shall be registered and shall be guaranteed. Guarantees shall be made on the elemental basis. Sources of the elements guaranteed and proof of availability shall be provided the .... upon request. The minimum percentages which will be accepted for registration are as follows:

Element	%
Calcium (Ca)	1.00
Magnesium (Mg)	0.50
Sulfur (S)	1.00

Element	%
Boron (B)	0.02
Chlorine (Cl)	0.10
Cobalt (Co)	0.0005
Copper (Cu)	0.05
Iron (Fe)	0.10
Manganese (Mn)	0.05
Molybdenum (Mo)	0.0005
Sodium (Na)	0.10
Zinc (Zn)	0.05

Guarantees or claims for the above listed plant nutrients are the only ones which will be accepted. Proposed labels and directions for use of the fertilizer shall be furnished with the application for registration upon request. Any of the above listed elements which are guaranteed shall appear in the order listed immediately following guarantees for the primary nutrients of nitrogen, phosphorus and potassium.

A warning or caution statement is required on the label for any product which contains 0.03% or more of boron in water soluble form. This statement shall carry the word "WARNING" or "CAUTION" conspicuously displayed, shall state the crop(s) for which the fertilizer is to be used, and state that the use of the fertilizer on any other than those recommended may result in serious injury to the crop(s).

Products containing 0.001% or more of molybdenum also require a warning statement on the label. This shall include the word "WARNING" or "CAUTION" and the statement that the application of fertilizers containing molybdenum may result in forage crops containing levels of molybdenum which are toxic to ruminant animals.

Examples of Warning or Caution Statements

Boron:
1. Directions: Apply this fertilizer at a minimum rate of 350 pounds per acre for Alfalfa or Red Clover seed production. CAUTION: Do not use on other crops. The boron may cause injury to them.
2. CAUTION: Apply this fertilizer at a maximum rate of 700 pounds per acre for Alfalfa or Red Clover seed production. Do not use on other crops; the boron may cause serious injury to them.
3. WARNING: This fertilizer carries added borax and is intended for use only on alfalfa. Its use on any other crops or under conditions other than those recommended may result in serious injury to the crops.

Molybdenum:
1. CAUTION: This fertilizer is to be used only on soil which responds to molybdenum. Crops high in molybdenum are toxic to grazing animals (ruminants).

2. Specialty Fertilizer Labels.

   The following information, if not appearing on the face or display side in a *readable and conspicuous form*, shall occupy at least the upper third of a side of the container and shall be considered the label.

   A.   Net Weight
   B.   Brand and Grade
   C.   Guaranteed Analysis:

   Total Nitrogen (N) ....   _____ %
   _____ % Ammoniacal Nitrogen †
   _____ % Nitrate Nitrogen †
   _____ % Water Insoluble Nitrogen ‡
   Available Phosphoric Acid (P_2O_5) ....   _____ %
   Soluble Potash (K_2O) ....   _____ %
   Additional Plant Nutrients as Prescribed by regulation. Sources of nutrients, when shown on the label, shall be listed below the guaranteed analysis.
   Potential Acidity or Basicity † ....   _____ lbs.
   Calcium Carbonate Equivalent per ton.

   D.   Name and address of registrant

   ....

3. Slowly Available Plant Nutrients.

   A.   No fertilizer label shall bear a statement that connotes or infers the presence of a slowly available plant nutrient, unless the nutrient or nutrients are identified.

   B.   When a fertilizer label infers or connotes that the nitrogen is slowly available through use of organic, organic nitrogen, urea-form, long lasting or similar terms, the guaranteed analysis must indicate the percentage of water insoluble nitrogen in the material, except manipulated animal and vegetable manures distributed as such and not mixed with other materials. When the water insoluble nitrogen is less than 15% of the total nitrogen, the label shall bear no reference to such designations.

   C.   To supplement B, it should be established that if a label states the amount of organic nitrogen present in a phrase, such as "Nitrogen in organic form equivalent to x% N," then the water insoluble nitrogen guarantee must not be less than 60% of the nitrogen so designated. For example, if the total nitrogen guarantee for a fertilizer is 10% and the label states, "Nitrogen in organic form equivalent to 2.5% N," the WIN guarantee must not be less than 1.5%, determined in the following manner: 2.5% multiplied by 0.6 to obtain 1.5%.

---

† If claimed or required.
‡ If claimed or the statement "organic" or "slow acting nitrogen" is used on the label.

D. The term "Coated-Slow Release" may be accepted as descriptive of products.

E. Further, the above phrase (D) be allowed for any products that can show a testing program substantiating the claim. (Testing under guidance of Experiment Station personnel, or a recognized reputable researcher, etc.) Water insoluble nitrogen must be guaranteed at the 15% of total nitrogen level as in organic materials.

F. AOAC method 2.064 (12th edition), or as it shall be designated in subsequent AOAC editions, is to be used to confirm the water insoluble nitrogen of coated products and others whose slow release characteristics depend on particle size. AOAC method 2.062 (12th edition) shall be used to determine the water insoluble nitrogen of other products applicable for these procedures.

4. Definitions.

Except as the .... designates otherwise in specific cases, the names and definitions for commercial fertilizers shall be those adopted by the Association of American Plant Food Control Officials.

5. Percentages.

The term of "percentage," by symbol or word, when used on a fertilizer label shall represent only the amount of individual plant nutrients or other factors in relation to the total product by weight.

6. Investigational Allowances.

A. A commercial fertilizer shall be deemed deficient if the analysis of nutrient is below the guarantee by an amount exceeding the values in the following schedule, or if the overall index value of the fertilizer is below 98%.§

Guarantee (%)	Nitrogen (%)	Available Phosphoric Acid (%)	Potash (%)
4 or less	0.49	0.67	0.41
5	0.51	0.67	0.43
6	0.52	0.67	0.47
7	0.54	0.68	0.53
8	0.55	0.68	0.60
9	0.57	0.68	0.65

§ For these investigational allowances to be applicable, the recommended AOAC procedures for obtaining samples, sample preparation and analysis must be used. These are described in Official Methods of Analysis of the Association of Official Analytical Chemists, 11th edition, 1970, and in succeeding issues of the Journal of the Association of Official Analytical Chemists. In evaluating replicate data, Table 19, page 935, Journal of the Association of Official Analytical Chemists, Vol. 49, No. 5, October 1966, should be followed.

Guarantee (%)	Nitrogen (%)	Available Phosphoric Acid (%)	Potash (%)
10	0.58	0.69	0.70
12	0.61	0.69	0.79
14	0.63	0.70	0.87
16	0.67	0.70	0.94
18	0.70	0.71	1.01
20	0.73	0.72	1.08
22	0.75	0.72	1.15
24	0.78	0.73	1.21
26	0.81	0.73	1.27
28	0.83	0.74	1.33
30	0.86	0.75	1.39
32 or more	0.88	0.76	1.44

For guarantees not listed, calculate the appropriate value by interpolation.

Averaging at least two values must be adhered to. If more than two values are obtained, all significant values must be averaged. Values carried to two decimals are needed in applying averages to this table. Values may be "rounded" to one place where preferred in reporting.

The overall index value is calculated by comparing the commercial value guaranteed with the commercial value found. Unit values of the nutrients used shall be those referred to in Section 10.

Overall index value—example of calculation for a 10-10-10 grade found to contain 10.1% Total Nitrogen (N), 10.2% Available Phosphoric Acid (P_2O_5), and 10.1% Soluble Potash (K_2O). Nutrient unit values are assumed to be $3 per unit N, $2 per unit P_2O_5, and $1 per unit K_2O.

10.0 units N	$\times$ 3 =	30.0
10.0 units P_2O_5	$\times$ 2 =	20.0
10.0 units K_2O	$\times$ 1 =	10.0
Commercial value guarantee	=	60.0
10.1 units N	$\times$ 3 =	30.3
10.2 units P_2O_5	$\times$ 2 =	20.4
10.1 units K_2O	$\times$ 1 =	10.1
Commercial value found	=	60.8

$$\text{Overall index value} = \frac{60.8}{60.0} \times 100 = 101.3\%$$

B.   Secondary and minor elements shall be deemed deficient if any element is below the guarantee by an amount exceeding the values in the following schedule:

Element	Allowable Deficiency
Calcium	0.2 unit + 5% of guarantee
Magnesium	" " "
Sulfur	" " "
Boron	0.003 unit + 15% of guarantee
Cobalt	0.0001 unit + 30% of guarantee
Molybdenum	" " "
Chlorine	0.005 unit + 10% of guarantee
Copper	" " "
Iron	" " "
Manganese	" " "
Sodium	" " "
Zinc	" " "

The maximum allowance when calculated in accordance to the above shall be 1 unit (1%).

7. Sampling.

Sampling equipment and procedures shall be those adopted by the Association of Official Analytical Chemists wherever applicable.

8. Breakdown of Plant Food Elements Within the Guaranteed Analysis.

When a plant nutrient guarantee is broken down into the component forms, the percentage for each component shall be shown before the name of the form. EXAMPLE: 4% Nitrate Nitrogen.

Chapter 14

AMENDING THE PHYSICAL PROPERTIES
OF SOILS FOR
PLANTING AND POTTING PURPOSES

For centuries, cover crops, crop residues, animal manures and other organic materials have been applied to the soil to help maintain or improve soil physical properties and productivity. One of the earliest recorded instances of adding manure to the soil was in 900 B.C., as recorded in the Greek epic poem, *The Odyssey*.

Fig. 14-1. Composting of municipal sludge.

From that time to the present, the physical condition of soils has been improved by the addition of soil amendments. As we continue to work our lands and to grow and harvest plants, the need for soil improvement will always be present. This is particularly true today as urbanization forces us to use less than ideal soils.

The availability and types of amendments to improve the physical properties of soils continually change. Often these materials are by-products. Organic wastes such as bark, sawdust, manure, sludge and mushroom composts are examples. Supplies of rice hulls, cedar and other wood residue are increasing, while redwood and peat moss supplies are declining. Other materials will undoubtedly come into use in the future.

DEFINITION OF PHYSICAL SOIL AMENDMENTS

For the purposes of this chapter, we have defined "physical soil amendments" as any substances used for the purpose of promoting plant growth or improving the quality of crops by conditioning soils solely through physical means. This centers primarily on soil improvement by improving water retention and permeability.

NEED FOR PHYSICAL SOIL AMENDMENTS

When a soil is adequately supplied with sufficient levels of air, water, organic matter and nutrients, little amending is necessary. Such a soil is at a premium. Generally, it is necessary to add a soil amendment to improve the physical condition of the soil so that proper conditions can be obtained and maintained for good plant growth.

If a grower chooses to replace the existing soil with a good top-soil, this is usually more costly than amending. Also, unless careful measures are followed, soil layering often results, which causes problems with air, water and root movement.

When one plants trees and shrubs, selecting a proper amendment is very important. Plants are expected to beautify an area over extended periods of time. Since it is impractical to amend a soil once an area has been planted, this should be done before plants are established.

The soil is very similar to the foundation of a building. The question to be asked is "Will this soil support good plant growth?" If not, it will need to be amended. Limited budgets are not an adequate excuse. It is much wiser to spend the money on amendments

Fig. 14-2. Calendulas grown in amended soil on left and unamended soil on right.

and then use less expensive, smaller or even seeded plants in planting. Rather than set a five-dollar size plant into a fifty-cent hole, it is wiser to set a fifty-cent size plant into a five-dollar hole.

SOIL PHYSICAL PROPERTIES

The physical properties of soils are related to soil textures as illustrated in Table 14-1.

Table 14-1. Soil Physical Properties Related to Soil Texture

Soil Texture	Permeability	Water Retention
Sand	High	Low
Loam	Medium	Medium
Silt	Low	High
Clay	Low	High

Sandy soils have properties that are almost directly opposite those of clay soils. Permeability or porosity is high in a sandy soil and low in a clay soil. These differences are recognized in maintenance practices such as irrigation, but all too often, they are overlooked at the time of planting and amending. For optimum results, it is essential that soil texture be considered when the choice of amendments is made.

LONGEVITY OF AMENDMENTS

None of the organic compounds of plant residues are indestructible, although some are very resistant to decomposition. Considering amendments as a whole, their longevity in the soil depends primarily upon aeration, moisture content, temperature and available nutrients. The rate of decomposition is reduced when aeration is limited, under low or excessive moisture conditions, with low temperatures, and when nutrients, primarily nitrogen, are limited.

Within about six months most of the celluloses disappear, but the lignins may last in soil for years. Examples of some organic residues and their relative decomposition rates are given as follows:

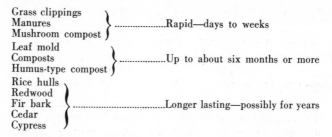

```
Grass clippings    ⎫
Manures            ⎬ ..................Rapid—days to weeks
Mushroom compost   ⎭

Leaf mold          ⎫
Composts           ⎬ ..................Up to about six months or more
Humus-type compost ⎭

Rice hulls         ⎫
Redwood            ⎪
Fir bark           ⎬ ...........................Longer lasting—possibly for years
Cedar              ⎪
Cypress            ⎭
```

TYPES OF PHYSICAL SOIL AMENDMENTS

Physical soil amendments are classified into two broad categories: organic and inorganic. Organic amendments are derived from living sources. They improve soil properties by physically separating the soil particles and by increasing nutrient and water absorptive capacity. The stable decomposition residue from organic amendments is humus.

Inorganic soil amendments do not contain humus nor do they contribute to the production of it. Instead, their principal function is to act as wedges in separating the soil particles physically. Some of these materials also aid in the retention of water.

Fig. 14-3. Bulk redwood soil amendment in storage.

A description of the more common organic and inorganic amendments follows.

Organic Amendments

Peat—Peat is the general classification of plant residues that have accumulated and undergone partial or incomplete decomposition in water or in excessively wet areas such as swamps and bogs. Peats and peat moss exhibit a considerable range in moisture-holding capacity, organic content and longevity. Peats can promote very heavy root growth under conditions of non-compaction, but when they are mixed in soils subject to compaction, the reverse is true. Once they become dry, peats are difficult to wet.

Peats are classified into different types, according to the kinds of plants from which they derived and the length of time they have been decomposing.

Moss peat (peat moss, sphagnum peat)—This partially decomposed peat is very low in plant nutrient content and has a low pH. It has a high water-holding capacity, equal to 15 to 30 times its own weight. Moss peat is lightweight, porous and somewhat difficult to mix into soils. It is light brown to tan in color. Moss peat decomposes at a moderate rate because of its high cellulose content.

Fibrous peat (reed-sedge)—This peat has the highest commercial value and is of the most interest to turfgrass and landscape people. It accumulates in swamps or bogs along the edges and beneath the moss layers in the bogs. Reddish to dark brown in color, this rather fibrous material has undergone longer decomposition and is resistant to further decay. It does not have as high a moisture-holding capacity as moss peat.

Wood sawdust and shavings—Various grades of wood residues are widely used in amending turfgrass and landscape soils and in preparing container soil mixes. Possessing only poor to fair water- and nutrient-holding capacities, these amendments provide good water infiltration and oxygen diffusion into soils.

The rate of decomposition varies with the kind of wood used. Wood residues from pine and fir may only have a residual of several months, while residues from redwoods, cedars and cypress may last up to five years. When they are incorporated into the soil, additional nitrogen needs to be added to prevent nitrogen starvation of plants growing in the amended soils.

Ground fir bark—This product has proven to be a valuable amendment in the turfgrass and landscape industry. Possessing only low nutrient-holding and fair water-holding capacity, it resists compaction well. The round shape of the bark provides good water and oxygen diffusion rates into the soil. Ground fir bark lasts about five years in the soil.

Shredded rice hulls—Rice hulls are derived from the seed coat of rice. They are excellent soil conditioners since they hold soils open and increase soil porosity, thereby improving water infiltration and oxygen diffusion rates. Rice hulls do not have a good water- nor a good nutrient-holding capacity. The hulls decay very slowly in the soil, lasting up to ten years.

Composts—Composts are made by decomposing plant residues. They usually have a good nutrient-holding capacity and are very active biologically. Well-decomposed composts aid in creating good, desirable soil structure. One special type is mushroom compost, a by-product of the mushroom growing beds, consisting of soil, straw, manure, wood or peat and agricultural minerals. It provides moderate nutrient and fair humus contribution to soils. Unless these materials are leached, salinity may be a problem with them.

Rotted manures—For use as soil amendments, steer and dairy manures are best used when they have been well aged, rotted and composted. This lessens the problem of possible high salt content, odors and weed seed contamination. The water- and nutrient-holding capacities of rotted manures are fair.

Digested sludges—These processed sludges from sewage are produced after undergoing about 14 days of aerobic (with air) and anerobic (without air) digestion. Afterwards, they are pumped into basins and allowed to dry. Unless they are modified by the addition of wood products, digested sludges are not generally suitable as a soil amendment. Perhaps their best value is realized when they are used as an organic base for fertilizers.

Composted sludges—Composted sludge is made by centrifuging after digestion to reduce water content, followed by windrowing with biologically active dried sludge and turned daily for about 40 days. The resulting well-decomposed product approaches the characteristics of humus. When used in silt and clay soils, it improves aggregation; when used in sandy soils, it improves moisture-holding capacity and biological activity.

Fig. 14-4. A sand golf green showing the value of an organic amendment (sludge) in the foreground versus pure sand in the background.

Inorganic Amendments

Calcined clay—This long lasting fired clay resists compaction very well and has good water infiltration and oxygen diffusion rates.

Water-holding capacity is high but water-releasing and nutrient-holding characteristics are poor. It makes a good surface mulch. In clay soils, it helps to develop very deep roots, but with only a sparse density. Costs for these materials are high, but they can aid in the establishment of high value turfgrass and landscape plantings. Soils with calcined clays mixed into them often have a hard surface.

Pumice—This amending material is a spongy, light, porous volcanic rock. It is relatively costly, but has good properties in helping to establish plantings. It is only fair as a surface mulch, but when mixed into the soil, it promotes good water infiltration and oxygen diffusion rates and a deep and dense root system. Water- and nutrient-holding properties of pumice are only poor to fair. Pumice from some sources may have excessively fine pores which results in poor water-releasing properties. This material is usually limited to greenhouse propagation mixes and specialized potting soils.

Vermiculite—Upon heating a naturally occurring micaceous mineral ore, the internal moisture turns to steam and pops the flakes to many times their original size. The resultant fluffy, lightweight product can hold water in amounts several times its own weight, although even when thoroughly wet, there is still sufficient air (oxygen) for plant roots to grow. Lasting indefinitely, vermiculite is used mostly in specialized greenhouse propagation and potting soil mixes.

Perlite—Perlite is a naturally occurring, glassy volcanic rock which is heat-treated like vermiculite to produce a whitish-colored, bead-like particle with very low bulk density. It is used as a soil amendment to provide good aeration and drainage. It is often used in combination with peat moss as a medium for the vegetative propagation of general ornamental nursery stock and as a lightweight ingredient in potting soils for both commercial and home use. Perlite is inert and water insoluble, and it has a pH of 6 to 7.

Sand—Although low in cost and with great longevity in the soil, sand has limited value as a soil amendment, except as used in special sand mixes such as golf and bowling greens and nursery production. Low in both water- and nutrient-holding capabilities, sand causes finer silt or clay soils to compact, unless used at very high rates, perhaps up to 80 percent of the mix.

SELECTING PHYSICAL SOIL AMENDMENTS FOR VARYING SOIL CONDITIONS

The following summary indicates the soil-amending properties of some commonly used amendments. No single one may be totally satisfactory, so a combination of amendments is often used. Sandy soils are benefited when amended with a humus type of material to increase water retention and nutrient-holding capacity. Clayey and silty soils, in contrast, need fibrous-type materials to increase permeability and water retention. But humus may be needed to give better water and nutrient retention if the soil is low in organic matter content.

Table 14-2.   Soil-Amending Properties of Some Materials

	Amendment	Permeability	Water Retention
Fibrous	Peat	Low-Medium	Very high
	Wood residues	High	Low-Medium
	Ground fir barks	High	Low-Medium
	Rice hulls	High	Low-Medium
Non-fibrous or humus	Composts	Low-Medium	Medium-High
	Rotted manures	Low-Medium	Medium
	Composted sludges	Low	High
Inorganic	Calcined clay	High	High
	Pumice	High	Low-Medium
	Vermiculite	High	High
	Perlite	High	Low
	Sand	High	Low

DETERMINING QUANTITIES OF ORGANIC AMENDMENTS

When choosing amendments, one must consider whether or not they will be effective, practical and economical, and one must also consider the use, size and value of the area to be amended. A good loam soil not subject to much compaction may not need amending. It is generally ineffective to amend a soil using less than 25 percent by volume of amendment. Heavy use, high value and fine textured soils with poor structure may benefit from being amended up to 40 percent or 50 percent by volume. Some general guidelines in determining the quantity of soil amendments needed for various soil textures are given in Table 14-3.

Turfgrass and Ground Covers

The general rule to amend soils for these plants is to use about

33 percent by volume. For heavily used areas and with fine textured silt and clay soils, this may be increased to 40 percent by volume. For good loam soils where little traffic is expected, this may be reduced to a minimum of 25 percent by volume. Table 14-4 may be used to determine the volume of soil amendment needed to amend the soil to various depths.

Bedding Plants, Ground Covers and Vegetable Gardens

These are usually high volume, low acreage areas which may be amended by adding 2 inches of amendment over a previously deeply cultivated soil. If the grower incorporates the amendment to a 6-inch depth, the soil will be amended 33 percent by volume.

Backfill Around Landscape Shrubs and Trees

Recommendations are based on the size of the planting hole. Table 14-5 indicates some hole sizes for various containers and the quantities of amendments to be added to provide either 33 percent or 50 percent by volume.

Container Mixes

Soil mixes for container plantings vary widely in composition, for they are dependent upon use and availability of materials. The following two examples are formulas which have given good success, although they are often modified.

Example 1—Landscape container mix

50%—sand or sandy loam
50%—fibrous organic material such as
redwood or rice hulls

plus, per each cubic yard add:
4 oz. potassium sulfate
6 oz. potassium nitrate
2½# single superphosphate
3½# dolomite lime
1½# calcium carbonate–lime
1¼# gypsum

Example 2—Greenhouse or foliage plant mix

33% sand
33% peat moss
33% redwood or perlite

Table 14-3. Amendment Needed, Based on Soil Texture

Texture	Percent of Amendment
Sand	35
Loamy sand, Sandy loam	30
Sandy clay loam, Sandy clay, Loam	25
Silt loam, Silty clay loam, Clay loam	30
Silt, Silty clay, Clay	35

Table 14-4. Determining the Volume of Soil Amendments to Add for Various Depths of Treatment (cubic yards per 1,000 square feet)*

Percent of Amendment	Depth of Amended Soil in Inches (Soil and Amendments)						
	3 in.	4 in.	5 in.	6 in.	7 in.	8 in.	9 in.
1	0.09	0.12	0.15	0.18	0.21	0.25	0.28
2	0.18	0.25	0.31	0.37	0.43	0.49	0.55
3	0.28	0.37	0.46	0.56	0.65	0.74	0.83
4	0.37	0.49	0.62	0.74	0.86	0.99	1.11
5	0.46	0.61	0.77	0.93	1.08	1.23	1.39
10	0.93	1.23	1.54	1.85	2.16	2.47	2.78
15	1.39	1.85	2.32	2.78	3.24	3.70	4.17
20	1.85	2.47	3.09	3.71	4.32	4.94	5.55
25	2.32	3.08	3.86	4.63	5.40	6.17	6.95
30	2.78	3.70	4.64	5.56	6.48	7.41	8.33
35	3.24	4.32	5.40	6.48	7.57	8.64	9.72
40	3.70	4.94	6.18	7.41	8.64	9.88	11.13
45	4.17	5.55	6.95	8.33	9.72	11.10	12.52
50	4.63	6.17	7.72	9.26	10.80	12.34	13.88

* These values are additive; i.e., to calculate values not in the table, simply add the percentage figures. For example, to calculate 33% at a 6-inch depth, add the 30% figure and the 3% figure (5.56 + 0.56 = 6.12).

Table 14-5. Determining the Volume of Soil Amendments (cubic feet) Required to Amend Backfill Soils by 33% or 50% for Various-Sized Containers (cans or boxes)

			Container Size									
	1 Gal.		5 Gal.		7 Gal.		15 Gal.		20 In.		24 In.	
Hole Size*	33%	50%	33%	50%	33%	50%	33%	50%	33%	50%	33%	50%
2W, D	0.1	0.2	0.8	1.2	1.3	2.0	1.6	2.4	3.8	5.7	6.7	10
2W, D + ½ foot	0.3	0.5	1.3	2.0	2.0	3.0	2.6	4.0	5.4	8.2	9.2	14
2W, D + 1 foot	0.4	0.6	1.8	2.7	2.7	4.1	3.5	5.3	6.3	11.0	12.0	17
2W, 2D	0.3	0.5	1.8	2.7	2.9	4.4	3.6	5.5	8.8	13.0	16.0	24

	30 In.		36 In.		42 In.		48 In.		54 In.		60 In.	
Hole Size*	33%	50%	33%	50%	33%	50%	33%	50%	33%	50%	33%	50%
2W, D	14	22	23	34	30	45	44	67	52	79	73	111
2W, D + ½ foot	18	27	28	43	37	56	54	82	64	97	87	132
2W, D + 1 foot	23	39	34	51	44	67	64	97	75	114	101	154
2W, 2D	33	50	53	80	69	105	103	156	121	184	172	260

* W = width of container; D = depth of container.

plus, per each cubic yard add:
2# single superphosphate
4 oz. potassium sulfate
4 oz. potassium nitrate
5# hoof and horn meal or blood meal

INCORPORATION OF SOIL AMENDMENTS

A uniform soil mixture must be provided throughout the entire root zone. Poor mixing of materials will produce layers or pockets, resulting in interfaces that cause problems with air, water, fertilizer and root movement through the soil.

With the exception of potting soils or specialty growing mediums, physical soil amendments are mixed with the existing site soil. This is described as on-site mixing. On-site mixing involves the uniform distribution of the desired materials in layers over the surface of the soil followed by thorough mixing with some type of cultivating equipment, ensuring that no layers or pockets of materials remain. Roto-tillers, turning cultivators, disks and harrows are often used for mixing, although disks tend to leave pockets at turns and high speed tillers may separate soil particles too finely.

Fig. 14-5. Bulk organic soil amendments being mixed for container plants.

After cultivation, the area should be graded to provide for positive drainage, while ensuring that no low spots or pockets remain.

Mulches

Mulches are those substances that are spread upon the ground to:
1. Protect against soil erosion.
2. Protect the roots of plants from heat, cold or drought.
3. Keep fruit clean.
4. Assist in providing uniform seed or plant germination and establishment.

Fig. 14-6. Nurseryman inspecting root growth of plants grown in prepared organic medium.

Not only do mulches control soil erosion that causes gullies and rills, but they also protect against sheet erosion caused by the action of splashing raindrops. They further prevent the displacement of seeds and fertilizers and can aid in minimizing weeds and surface soil crusting. Most commonly, mulches are used to retain soil moisture.

Commonly used mulch materials are straw, plastic, wood products such as barks or chips, sludge and fibrous cellulose materials such as those used in hydroseeding. Combinations of these products are also used.

Mulching and Top Dressing

Generally, mulching involves the spreading of materials at depths of ½ inch or greater. Top dressing should be limited to depths of ⅛ inch to ¼ inch. Applying materials thicker than this after seeding may create germination problems or interfaces when used on lawns or ground covers, thus interfering with water movement and promoting fungal diseases. Also, on such areas it would be wise to use

Table 14-6.   Volume of Materials Required for Mulching or Top Dressing

	Approximate Volumes of Material	
Depth (in.)	Per 1,000 sq. ft. (cu. ft.)	Per Acre (cu. yds.)
⅛	10	16
¼	20	32
⅜	30	48
½	40	64
⅝	50	80
¾	60	96
⅞	70	112
1	80	128

the faster decomposing materials, for those with longer residual could contribute to thatch and eventual water infiltration problems.

By necessity, this chapter has been concerned only with the physical properties of soils and soil amendments. To explore this subject more fully, one should examine the supplementary reading references. Additionally, since the organic soil amendments have other chemical and biological values, one should read Chapters 1 and 8 for a more complete understanding of these materials.

SUPPLEMENTARY READING

1. *Amending Soils.* P. A. Rogers. California Landscape Management. March/April 1976.
2. "Evaluating Soil Amendments." J. E. Warneke and S. J. Richards. *California Agriculture.* September 1974.
3. "Peat Classifications." J. R. Watson. *California Turfgrass Culture.* July 1967; *Turfgrass Times.* 1968.
4. "Review of Soil Amendments." W. H. McNeely and W. C. Morgan. *Turfgrass Times.* March 1968.
5. "Soil Amendments, What Can They Really Do?" R. L. Branson. *Western Landscape News.* February 1979.
6. *Soil Fertility and Fertilizers*, Third Edition. S. L. Tisdale and W. L. Nelson. The Macmillan Company. 1974.
7. *Soil Preparation Specifications.* Kellogg Supply, Inc. 1978.
8. "The Use of Physical Soil Amendments in Turfgrass Management." W. C. Morgan, J. Letey, S. J. Richards and N. Valoras. *California Turfgrass Culture.* July 1968.

Appendix A

GLOSSARY OF TERMS

AAPFCO—American Association of Plant Food Control Officials.

ABSORPTION—The process by which a substance is taken into and included within another substance, e.g., intake of water by soil, or intake of gases, water, nutrients or other substances by plants.

ACID-FORMING—A term applied to any fertilizer that tends to make the soil more acid.

ACID SOIL—A soil with a pH value below 7.0. A soil having a preponderance of hydrogen over hydroxyl ions in the soil solution.

ACTIVATED SEWAGE SLUDGE—An organic fertilizer made from sewage freed from grit and coarse solids and aerated after being inoculated with microorganisms. The resulting flocculated organic matter is withdrawn from the tanks, filtered with or without the aid of coagulants, dried, ground and screened.

ADSORPTION—The increased concentration of molecules or ions at a surface, including exchangeable cations and anions on soil particles.

AERATION, SOIL—The exchange of air in soil with air from the atmosphere. The composition of the air in a well areated soil is similar to that in the atmosphere; in a poorly aerated soil, the air in the soil is considerably higher in carbon dioxide and lower in oxygen than the atmosphere above the soil.

AGGREGATE—A group of soil particles cohering so as to behave mechanically as a unit.

ALKALINE—A basic reaction in which the pH reading is above 7.0, as distinguished from acidic reaction in which the pH reading is below 7.0.

ALKALINE SOIL—A soil that has an alkaline reaction, i.e., a soil for which the pH reading of the saturated soil paste is above 7.0.

ALKALI SOIL—A soil that contains sufficient exchangeable sodium to interfere with the growth of most plants, either with or without appreciable quantities of soluble salts. (See also NONSALINE-ALKALI SOIL and SALINE-ALKALI SOIL.)

AMENDMENT—Any material, such as lime, gypsum, sawdust or synthetic conditioners, that is worked into the soil to make it more productive. Strictly, a fertilizer is also an amendment, but the term *amendment* is used more commonly for added materials other than fertilizer.

AMINO ACIDS—Nitrogen-containing organic compounds, large numbers of which link together in the formation of the protein molecule. Each amino acid molecule contains one or more amino ($-NH_2$) groups and at least one carboxyl ($-COOH$) group. In addition, some amino acids (cystine and methionine) contain sulfur.

AMMONIATED SUPERPHOSPHATE—A product formed by ammoniating superphosphate.

AMMONIATION—A process wherein ammonia (anhydrous, aqua or a solution containing ammonia and other forms of nitrogen) is used to treat superphosphate to form ammoniated superphosphate, or to treat a mixture of fertilizer ingredients (including phosphoric acid) in the manufacture of a multinutrient fertilizer.

AMMONIFICATION—Formation of ammonium compounds, or ammonia.

AMMONIUM CITRATE $(NH_4)_3C_6H_5O_7$—A salt formed from ammonia and citric acid. A neutral ammonium citrate solution, prepared by the official methods of the AOAC, is used as a reagent in the determination of "available" phosphoric acid in fertilizers. After a sample is washed with water to remove the water-soluble phosphoric acid (P_2O_5), the residue is treated with the neutral ammonium citrate solutions, as prescribed by the official methods, and the phosphoric acid removed by this extraction is termed "citrate-soluble." The sum of the water-soluble plus the citrate-soluble phosphoric acid is termed "available."

ANALYSIS—The percentage composition as found by chemical analysis, expressed in those terms that the law requires and permits. Although analysis and grade sometimes are used synonymously, the term grade is applied only to the three primary plant foods—nitrogen (N), available phosphate (P_2O_5) and potash (K_2O)—and is stated as the guaranteed minimum quantities present. (See also GRADE.)

ANGLE OF REPOSE—The angle between the horizontal and the slope of a pile of loose material at equilibrium.

ANION—An ion carrying a negative charge of electricity.

AOAC—Association of Official Analytical Chemists (of North America).

APATITE (rock phosphate)—A mineral phosphate having the type formula $Ca_{10}(X_2)(PO_4)_6$ where X is usually fluorine, chlorine or the hydroxyl group, either singly or together. Fluorapatite is widely distributed as the crystalline mineral and as amorphous phosphate rock, both forms of which are important fertilizer materials. Crystalline fluorapatite contains from 38.0 to 41.0 percent phosphoric acid (P_2O_5) and from 3.2 to 4.3 percent fluorine. Calcium hydroxyapatite or calcium hydroxy-phosphate, $Ca_{10}(OH)_2(PO_4)_6$, may be formed to a small extent in ammoniated superphosphate.

AVAILABLE—In general, a form capable of being assimilated by a growing plant. Available nitrogen is defined as the nitrogen that is water-soluble plus what can be made soluble or converted into free ammonia. Available phosphoric acid is that portion which is water-soluble plus the part which is soluble in ammonium citrate. Available potash is defined as that portion soluble in water or in a solution of ammonium oxalate.

AVAILABLE NUTRIENT IN SOILS—The part of the supply of a plant nutrient in the soil that can be taken up by plants at rates and in amounts significant to plant growth.

AVAILABLE WATER IN SOILS—The part of the water in the soil that can be taken up by plants at rates significant to their growth; usable; obtainable.

BASE EXCHANGE—The replacement of cations, held on the soil complex, by other cations. (See also CATION EXCHANGE CAPACITY.)

BASIC SLAG—A by-product in the manufacture of steel, containing lime, phosphate and small amounts of other plant food elements such as sulfur, manganese and iron. Basic slags may contain from 10 to 17 percent phosphate (P_2O_5) and 35 to 50 percent calcium oxide (CaO) and 2 to 10 percent mangnesium oxide (MgO). The available phosphate content of most American slag is in the range of 8 to 10 percent.

BONE MEAL—Raw bone meal is cooked bones ground to a meal without any of the gelatine or glue removed. Steamed bone meal has been steamed under pressure to dissolve out part of the gelatine.

BRAND—The trade name assigned by a manufacturer to his particular fertilizer product.

BRIMSTONE—Sulfur.

BUFFER CAPACITY OF SOILS—The ability of the soil to resist a change in its pH (hydrogen ion concentration) when acid-forming or base-forming materials are added to the soil.

BULK BLENDING—The practice of mixing dry, individual, granular materials or granulated bases. The product is a mixture of granular materials rather than a granulated mixture.

BULK DENSITY—The ratio of the mass of water-free soil to its bulk volume. Bulk density is expressed in pounds per cubic foot or grams per cubic centimeter and is sometimes referred to as "apparent density." When expressed in grams per cubic centimeter, bulk density is numerically equal to apparent specific gravity or volume weight.

CALCAREOUS SOIL—A soil containing calcium carbonate, or a soil alkaline in reaction because of the presence of calcium carbonate; a soil containing enough calcium carbonate to effervesce (fizz) when treated with dilute hydrochloric acid.

CALCIUM CARBONATE EQUIVALENT—The amount of calcium carbonate required to neutralize the acidity produced by a given quantity of fertilizer product.

CARBOHYDRATE—A compound containing carbon, hydrogen and oxygen. Usually the hydrogen and oxygen occur in the proportion of 2 to 1, such as in glucose ($C_6H_{12}O_6$).

CARBON: NITROGEN RATIO—The ratio obtained by dividing the percentage of organic carbon by percentage of nitrogen.

CATION—An ion carrying a positive charge of electricity. The common soil cations are calcium, magnesium, sodium, potassium and hydrogen.

CATION EXCHANGE CAPACITY—The total quantity of cations which a soil can adsorb by cation exchange, usually expressed as milliequivalents per 100 grams. Measured values of cation exchange capacity depend somewhat on the method used for the determination.

CHELATES—Certain organic chemicals, known as chelating agents, form ring compounds in which a polyvalent metal is held between two or more atoms. Such rings are chelates. Among the best chelating agents known are ethylenediaminetetraacetic acid (EDTA), hydroxyethylenediaminetriacetic acid (HEDTA) and diethylenetriaminepentaacetic acid (DTPA).

CHLOROSIS—Yellowing of green portions of a plant, particularly the leaves.

CITRATE-SOLUBLE PHOSPHORIC ACID—That fraction of the phosphoric acid insoluble in water but soluble in neutral ammonium citrate. However, since that soluble in water is also soluble in ammonium citrate, "citrate-soluble" may be used to indicate the sum of water-soluble plus citrate-soluble phosphoric acid. (See also AVAILABLE.)

CLAY—A minute soil particle less than 0.002 millimeters in diameter.

COLLOID—The soil particles (inorganic or organic) having small diameters ranging from 0.20 to 0.005 microns. Colloids are characterized by high base exchange.

COMPLETE FERTILIZER—A fertilizer containing all three of the primary fertilizer nutrients (nitrogen, phosphate and potash) in sufficient amounts to be of value as nutrients.

CONDITIONER (of fertilizer)—A material added to a fertilizer to prevent caking and to keep it free-flowing.

CONDUCTIVITY, ELECTRICAL—A physical quantity that measures the readiness with which a medium transmits electricity. Commonly used for expressing the salinity of irrigation waters and soil extracts because it can be directly related to salt concentration. It is expressed in siemens and millisiemens (ms) per centimeter (or mhos, millimhos and micromhos per centimeter) at $25°$ C.

CURING—The process by which superphosphate or mixed fertilizers are stored until the chemical reactions have run to, or nearly to, completion.

CYTOPLASM—The portion of the protoplasm of a cell outside the nucleus.

DAMPING-OFF—Sudden wilting and death of seedling plants resulting from attack by microorganisms.

DENITRIFICATION—The process by which nitrates or nitrites in the soil or organic deposits are reduced to lower oxides of nitrogen by bacterial action. The process results in the escape of nitrogen into the air.

DOLOMITE—A material used for liming soils in areas where magnesium as well as calcium is needed. Made by grinding dolomitic limestone, which contains both magnesium carbonate, $MgCO_3$, and calcium carbonate, $CaCO_3$. (See also LIME.)

ECOLOGY—The branch of biology that deals with the mutual relations among organisms and between organisms and their environment.

ELEMENTAL GUARANTEES—See GUARANTEES.

ENVIRONMENT—All external conditions that may act upon an organism or soil to influence its development, including sunlight, temperature, moisture and other organisms.

ENZYMES—Substances produced by living cells which can change the rate of chemical reactions. They are organic catalysts.

EROSION—The wearing away of the land surface by detachment and transport of soil and rock materials through the action of moving water, wind or other geological agents.

EVAPOTRANSPIRATION—The loss of water from a soil by evaporation and plant transpiration.

EXCHANGEABLE IONS—Ions held on the soil complex that may be replaced by other ions. There are ions held so tightly that they cannot be exchanged which are called nonexchangeable.

EXCHANGEABLE SODIUM PERCENTAGE—The degree of saturation of the soil exchange complex with sodium. It may be calculated by the formula:

$$\text{ESP} = \frac{\text{Exchangeable sodium (me/100 gm soil)}}{\text{Cation exchange capacity (me/100 gm soil)}} \times 100$$

FALLOW—Cropland left idle in order to restore productivity, mainly through accumulation of water, nutrients or both. Summer fallow is a common stage before cereal grain in regions of limited rainfall. The soil is tilled for at least one growing season to control weeds, to aid decomposition of plant residues and to encourage the storage of moisture for the succeeding grain crop. Bush or forest fallow is a rest period under woody vegetation between crops.

FERTILIZER—Any natural or manufactured material added to the soil in order to supply one or more plant nutrients. The term is generally applied to largely inorganic materials other than lime or gypsum sold in the trade.

FERTILIZER FORMULA—The quantity and grade of materials used in making a fertilizer mixture.

FERTILIZER GRADE—An expression that indicates the percentage of plant nutrients in a fertilizer. Thus a 10-20-10 grade contains 10 percent nitrogen (N), 20 percent phosphoric acid (P_2O_5) and 10 percent potash (K_2O).

FERTILIZER RATIO—The relative proportions of primary nutrients in a fertilizer grade divided by the highest common divisor for that grade; e.g., grades 10-6-4 and 20-12-8 have the ratio 5-3-2.

FIELD MOISTURE CAPACITY—The moisture content of soil in the field two or three days after a thorough wetting of the soil profile by rain or irrigation water. Field capacity is expressed as moisture percentage, dry-weight basis.

FIFTEEN-ATMOSPHERE PERCENTAGE—The moisture percentage, dry-weight basis, of a soil sample which has been wetted and brought to equilibrium in a pressure-membrane apparatus at a pressure of 221 psi. This characteristic moisture value for soils approximates the lower limit of water available for plant growth.

FIXATION—The process by which available plant nutrients are rendered unavailable or "fixed" in the soil. Generally, the process by which potassium, phosphorus and ammonia are rendered unavailable in the soil. Also the process by which free nitrogen is chemically combined either in nature or synthetically. (See also REVERSION and NITROGEN FIXATION.)

FORAGE—Unharvested plant material which can be used as feed by domestic animals. Forage may be grazed or cut for hay.

GRADE—The guaranteed analysis of a fertilizer containing one or more of the primary plant nutrient elements. Grades are stated in terms of the guaranteed percentages of nitrogen (N), available phosphate (P_2O_5) and potash (K_2O) in that order. For example, a 10-10-10 grade would contain 10 percent nitrogen, 10 percent available phosphate, and 10 percent potash. (See also ANALYSIS.)

GUANO—The decomposed dried excrement of birds and bats, used for fertilizer purposes. The most commonly known guano comes from islands off the coast of Peru and is derived from the excrement of sea fowl. It is high in nitrogen and phosphate, and at one time was a major fertilizer in this country.

GUARANTEES—The AAPFCO official regulation follows: The statement of guarantees of mixed fertilizer shall be given in whole numbers. All fertilizer components with the exception of potash (K_2O) and phosphoric acid (P_2O_5), if guaranteed, shall be stated in terms of the elements.

GYPSUM ($CaSO_4 \cdot 2H_2O$)—The common name for calcium sulfate, a mineral used in the fertilizer industry as a source of calcium and sulfur. Gypsum also is used widely in reclaiming alkali soils in the western United States. Gypsum cannot be used as a liming material, but it may reduce the alkalinity of sodic soils by replacing sodium with calcium. Another common name is landplaster. When pure it contains approximately 18.6 percent sulfur.

HARDPAN—A hardened or cemented soil horizon or layer. The soil material may be sandy or clayey and may be cemented by iron oxide, silica, calcium carbonate or other substances.

HORIZON, SOIL—A layer of soil, approximately parallel to the soil surface, with distinct characteristics produced by soil-forming processes.

HUMUS—The well-decomposed, more or less stable part of the organic matter in mineral soils.

HYDROGEN ION CONCENTRATION—(See pH.)

HYGROSCOPIC—Capable of taking up moisture from the air.

INORGANIC—Substances occurring as minerals in nature or obtainable from them by chemical means. Refers to all matter except the compounds of carbon, but includes carbonates.

INSOLUBLE—Not soluble. As applied to phosphoric acid in fertilizer, that portion of the total phosphoric acid which is neither soluble in water nor in neutral ammonium citrate. As applied to potash and nitrogen, not soluble in water.

ION—An electrically charged particle. As used in soils, an ion refers to an electrically charged element or combination of elements resulting from the breaking up of an electrolyte in solution. Since most soil solutions are very dilute, many of the salts exist as ions. For example, all or part of the potassium chloride (muriate of potash) in most soils exists as potassium ions and chloride ions. The positively charged potassium ion is called a cation, and the negatively charged chloride ion is called an anion.

KELP—A species of seaweed sometimes harvested for fertilizer material. Dried kelp will usually contain 1.6 to 3.3 percent nitrogen, 1 to 2 percent P_2O_5 and 15 to 20 percent K_2O.

LEACHING—The removal of materials in solution by the passage of water through soil.

LEACHING REQUIREMENT—The fraction of the water entering the

soil that must pass through the root zone in order to prevent soil salinity from exceeding a specified value. Leaching requirement is used primarily under steady-state or long-time average conditions.

LIME—Generally the term "lime," or "agricultural lime," is applied to ground limestone (calcium carbonate), hydrated lime (calcium hydroxide) or burned lime (calcium oxide), with or without mixtures of magnesium carbonate, magnesium hydroxide or magnesium oxide, and materials such as basic slag, used as amendments to reduce the acidity of acid soils. In strict chemical terminology, lime refers to calcium oxide (CaO), but by an extension of meaning it is now used for all limestone-derived materials applied to neutralize acid soils.

LIME REQUIREMENT—The amount of standard ground limestone required to bring a 6.6-inch layer of an acre (about 2 million pounds in mineral soils) of acid soil to some specific lesser degree of acidity, usually to slightly or very slightly acid. In common practice, lime requirements are given in tons per acre of nearly pure limestone, ground finely enough so that all of it passes a 10-mesh screen and at least half of it passes a 100-mesh screen.

LIQUID FERTILIZER—A fluid in which the plant nutrients are in true solution.

LOAM—The textural class name for soil having a moderate amount of sand, silt and clay. Loam soils contain 7 to 27 percent clay, 28 to 50 percent silt, and less than 52 percent sand. (In the old literature, especially English literature, the term "loam" applied to mellow soils rich in organic matter, regardless of the texture. As used in the United States, the term refers only to the relative amounts of sand, silt and clay; loam soils may or may not be mellow.)

LUXURY CONSUMPTION—The uptake by a plant of an essential nutrient in amounts exceeding what it needs. Thus if potassium is abundant in the soil, alfalfa may take in more than is required.

MACRONUTRIENTS—Nutrients that plants require in relatively large amounts.

MANURE—Generally, the refuse from stables and barnyards, including both animal excreta and straw or other litter. In some other countries the term "manure" is used more broadly and includes both farmyard or animal manure and "chemical manures," for which the term "fertilizer" is nearly always used in the United States.

MARL—An earthy deposit, consisting mainly of calcium carbonate, commonly mixed with clay or other impurities. It is formed chiefly at the margins of fresh-water lakes. It is commonly used for liming acid soils.

MICRONUTRIENTS—Nutrients that plants need in only small, trace or minute amounts. Essential micronutrients are boron, chlorine, copper, iron, manganese, molybdenum and zinc.

MILLIEQUIVALENT or MILLIGRAM EQUIVALENT (me)—One-thousandth of an equivalent, and in the case of sodium chloride, would be .023 gram of sodium and .0355 gram of chloride in one liter of water.

MUCK—Highly decomposed organic soil material developed from peat. Generally, muck has a higher mineral or ash content than peat and is decomposed to the point that the original plant parts cannot be identified.

MURIATE OF POTASH—Potassium chloride.

NITRIFICATION—The formation of nitrates and nitrites from ammonia (or ammonium compounds), as in soils by microorganisms.

NITROGEN FIXATION—Generally, the conversion of free nitrogen to nitrogen compounds. Specifically in soils, the assimilation of free nitrogen from the soil air by soil organisms and the formation of nitrogen compounds that eventually become available to plants. The nitrogen-fixing organisms associated with legumes are called symbiotic; those not definitely associated with the higher plants are non-symbiotic or free-living.

NONSALINE-ALKALI SOIL—A soil which contains sufficient exchangeable sodium to interfere with the growth of most crop plants but does not contain appreciable quantities of soluble salts. The exchangeable sodium percentage is greater than 15, the conductivity of the saturation extract is less than 4 millisiemens per centimeter (at 25° C.), and the pH of the saturated soil usually ranges between 8.5 and 10.0.

NUTRIENT, PLANT—Any element taken in by a plant which is essential to its growth, and which is used by the plant in elaboration of its food and tissue.

ORGANIC—Compounds of carbon other than the inorganic carbonates.

ORGANIC SOIL—A general term applied to a soil or to a soil horizon that consists primarily of organic matter, such as peat soils, muck soils and peaty soil layers.

ORTHOPHOSPHATE—A salt of orthophosphoric acid such as ammonium, calcium or potassium phosphate. Each molecule contains a single atom of phosphorus.

ORTHOPHOSPHORIC ACID—H_3PO_4.

PARENT MATERIAL—The unconsolidated mass of rock material (or peat) from which the soil profile develops.

PARTICLE DENSITY—The average density of the soil particles. Particle density is usually expressed in grams per cubic centimeter and is sometimes referred to as "real density" or "grain density."

PARTS PER MILLION (ppm)—A notation for indicating small amounts of materials. The expression gives the number of units by weight of the substance per million weight units of oven-dry soil. The term may be used to express the number of weight units of a substance per million weight units of a solution. The approximate weight of soil is 2 million pounds per acre-six inches. Therefore, ppm x 2 equals pounds per acre-six inches of soil, or ppm x 4 equals pounds per acre-foot of soil.

PEAT—The AAPFCO has adopted as official the following definition: "Peat is partly decayed vegetable matter of natural occurrence. It is composed chiefly of organic matter that contains some nitrogen of low activity."

PERCOLATION—The downward movement of water through soil.

PERMANENT WILTING PERCENTAGE—The moisture percentage of soil at which plants wilt and fail to recover turgidity (15 atmospheres). It is usually determined with dwarf sunflowers. The expression has significance only for non-saline soils.

PERMEABILITY, SOIL—The quality of a soil horizon that enables water or air to move through it. It can be measured quantitatively in terms of rate of flow of water through a unit cross section in unit time under specified temperature and hydraulic conditions. Values for saturated soils usually are called hydraulic conductivity. The permeability of a soil is controlled by the least permeable horizon even though the others are permeable.

pH—A numerical designation of acidity and alkalinity as in soils and other biological systems. Technically, pH is the common logarithm of the reciprocal of the hydrogen ion concentration of a solution. A pH of 7.0 indicates precise neutrality, higher values indicate increasing alkalinity, and lower values indicate increasing acidity.

PHOSPHATE—A salt of phosphoric acid made by combining phosphoric acid with ions such as ammonium, calcium, potassium or sodium.

PHOSPHATE ROCK—Phosphate-bearing ore composed largely of tricalcium phosphate. Phosphate rock can be treated with strong acids or heat to make available forms of phosphate. Finely ground rock phosphate is sometimes used in long-time fertility programs.

PHOSPHORIC ACID—A term that refers to the phosphorus content of a fertilizer, expressed as phosphoric acid (P_2O_5). The AAPFCO has adopted as official the following definition: "The term phosphoric acid designates P_2O_5." Phosphoric acid also refers to the acid H_3PO_4.

PHOTOSYNTHESIS—The process of conversion by plants of water and carbon dioxide into carbohydrates under the action of light. Chlorophyll is required for the conversion of light energy into chemical forms.

POLYPHOSPHATE—A salt of polyphosphoric acid such as ammonium, calcium or potassium polyphosphate. *Poly* means many and refers to multiple linkages of phosphorus in each molecule.

POLYPHOSPHORIC ACID—Condensed phosphoric acid ranging in P_2O_5 content from 68 to 83 percent.

POROSITY—The fraction of soil volume not occupied by soil particles.

POTASH—The AAPFCO has adopted as official the following definition: "The term potash designates potassium oxide (K_2O)."

PRIMARY PLANT NUTRIENTS (plant foods)—Nitrogen, phosphate (P_2O_5) and potash (K_2O).

PRODUCTIVITY—In simplest terms, the ability of the soil to produce. It differs from fertility to the extent that a soil may be fertile and yet unable to produce because of other limiting factors.

PROFILE, SOIL—A vertical section of the soil extending through all its horizons and into the parent material.

PROTEIN—Any of a group of nitrogen-containing compounds that yield amino acids on hydrolysis and have high molecular weights. Protein is an essential part of living matter and is one of the essential food substances of animals.

PROTOPLASM—The basic, jellylike substance in plant and animal cells; it is basic to all life processes.

PUDDLED SOIL—Dense, massive soil artificially compacted when wet and having no regular structure. The condition commonly results from the tillage of a clayey soil when it is wet.

PYRITE (FeS_2)—A mineral composed principally of iron and sulfur, with varying small amounts of other metals.

QUICK TESTS—Simple and rapid chemical tests of soils designed to give an approximation of the nutrients available to plants.

RATIO—(See FERTILIZER RATIO.)

RECLAMATION—The process of removing excess soluble salts or excess exchangeable sodium from soils and restoring lands to productivity.

REVERSION—The interaction of a plant nutrient with the soil which causes the nutrient to become less available. In fertilizer manufacturing, the excessive use of ammonia in ammoniation of phosphates results in phosphate reversion. (See also FIXATION.)

SALINE-ALKALI SOIL—A soil containing sufficient exchangeable sodium to interfere with the growth of most crop plants and containing appreciable quantities of soluble salts. The exchangeable sodium percentage is greater than 15, and the electrical conductivity of the saturation extract is greater than 4 ms per centimeter (at 25° C). The pH reading of the saturated soil is usually less than 8.5.

SALINE SOIL—A soil containing enough soluble salts to impair its productivity for plants but not containing an excess of exchangeable sodium.

SALT INDEX—An index used to compare solubilities of chemical compounds. Most nitrogen and potash compounds have high indexes, and phosphate compounds have low indexes. When applied too close to seed or on foliage, the compounds with high indexes cause plants to wilt or die.

SALTS—The products, other than water, of the reaction of an acid with a base. Salts commonly found in soils break up into cations (sodium, calcium, etc.) and anions (chloride, sulfate, etc.) when dissolved in water.

SAND—Individual rock or mineral fragments in soils having diameters ranging from .05 millimeters to 2.0 millimeters. Usually sand grains consist chiefly of quartz, but they may be of any mineral composition. The textural class name of any soil that contains 85 percent or more sand and not more than 10 percent clay.

SATURATED SOIL PASTE—A particular mixture of soil and water commonly used for measurements and for obtaining soil extracts. At saturation the soil paste glistens as it reflects light, flows slightly when the container is tipped and slides freely and cleanly from a spatula for all soils except those with high clay content.

SATURATION EXTRACT—The solution extracted from a soil at its saturation percentage.

SATURATION PERCENTAGE—The moisture percentage of a saturated soil paste, expressed on a dry-weight basis.

SECONDARY PLANT NUTRIENTS—Calcium, magnesium and sulfur.

SEPARATE, SOIL—One of the individual-size groups of mineral soil particles—sand, silt or clay.

SERIES, SOIL—A group of soils that have soil horizons similar in their differentiating characteristics and arrangement in the soil profile, except for the texture of the surface soil, and are formed from a particular type of parent material. Soil series is an important category in detailed soil classification. Individual series are given proper names from place names near the first recorded occurrence. Thus, names like Yolo, Panoche, Hanford and San Joaquin are names of soil series that appear on soil maps, and each connotes a unique combination of many soil characteristics.

SEWAGE SLUDGE—An organic product resulting from the treatment of sewage. The composition varies widely depending on the method of treatment.

SILT—(1) Individual mineral particles of soil that range in diameter between the upper size of clay, 0.002 mm, and the lower size of very fine sand, 0.05 mm. (2) Soil of the textural class silt containing 80 percent or more silt and less than 12 percent clay. (3) Sediments deposited from water in which the individual grains are approximately the size of silt, although the term is sometimes applied loosely to sediments containing considerable sand and clay.

SLURRY FERTILIZER—A fluid mixture containing dissolved and undissolved plant nutrient materials which requires continuous mechanical agitation to assure homogeneity.

SODIUM ADSORPTION RATIO—A ratio for soil extracts and irrigation waters used to express the relative activity of sodium ions in exchange reactions with soil.

$$SAR = \frac{Na^+}{\sqrt{(Ca^{++} + Mg^{++})/2}}$$

The ionic concentrations are expressed in milliequivalents per liter.

SODIUM PERCENTAGE—The percent sodium of total cations. Calculations are based on milliequivalents rather than weight.

SOIL MOISTURE STRESS—The sum of the soil moisture tension and the osmotic pressure of the soil solution. It is the force plants must overcome to withdraw moisture from the soil.

SOIL MOISTURE TENSION—The force by which moisture is held in the soil. It is a negative pressure and may be expressed in any convenient pressure unit. Tension does not include osmotic pressure values.

STRUCTURE, SOIL—The physical arrangement of the soil particles.

SUBSOIL—Roughly, that part of the soil below plow depth.

SUPERPHOSPHATE—The AAPFCO has adopted as official the following definition: "Superphosphate is a product obtained by mixing rock phosphate with either sulfuric acid or phosphoric acid or with both acids. (The grade that shows the available phosphoric acid shall be used as a prefix to the name. Example: 20 percent superphosphate.)"

SUPERPHOSPHORIC ACID—(See POLYPHOSPHORIC ACID.)

SUSPENSION FERTILIZER—A fluid containing dissolved and undissolved plant nutrients. The suspension of the undissolved plant nutrients may be inherent to the materials or produced with the aid of a suspending agent of non-fertilizer properties. Mechanical agitation may be necessary in some cases to facilitate uniform suspension of undissolved plant nutrients.

SYMBIOSIS—The living together of two different organisms with a resulting mutual benefit. A common example includes the association of rhizobia with legumes; the resulting nitrogen fixation is sometimes called symbiotic nitrogen fixation. Adjective: symbiotic.

TANKAGE—Dried animal residue. Process tankage is made from leather scrap, wool and other inert nitrogenous materials by steaming under pressure with or without addition of acid. This treatment increases the availability of the nitrogen to plants.

TENSIOMETER—A device used to measure the tension with which water is held in the soil.

TEXTURE, SOIL—The relative proportions of the various size groups of individual soil grains in a mass of soil. Specifically, it refers to the proportions of sand, silt and clay.

TILTH—The physical condition of a soil with respect to its fitness for the growth of plants.

TRACE ELEMENTS—(See MICRONUTRIENTS.)

TRANSPIRATION—Loss of water vapor from the leaves and stems of living plants to the atmosphere.

TRIPLE SUPERPHOSPHATE—A product that contains 40 to 50 percent available phosphoric acid. Triple superphosphate differs from ordinary superphosphate in that it contains very little calcium sulfate. In the fertilizer trade, the product is also called treble superphosphate, concentrated superphosphate, double superphosphate and multiple superphosphate.

UNIT—The AOAC has adopted as official the following definition: "A unit of plant food is twenty (20) pounds, or one percent (1 percent) of a ton."

VAPOR PRESSURE—The pressure exerted above a solution because of the tendency of vapor to escape from the liquid. Typical examples are the pressure above liquid anhydrous ammonia or ammonia-ammonium nitrate solutions. A negative vapor pressure value indicates that the vapor pressure above the liquid is less than atmospheric pressure. Vapor pressure is, of course, dependent upon the temperature at which it is measured. Increasing the temperature increases the vapor pressure above the liquid.

VOLATILIZATION—The evaporation or changing of a substance from liquid to vapor.

WATER TABLE—The upper surface of ground water.

WATER TABLE, PERCHED—The upper surface of a body of free ground water in a zone of saturation separated by unsaturated material from an underlying body of ground water.

WEATHERING—The physical and chemical disintegration and decomposition of parent material as in soil formation.

WILTING PERCENTAGE—(See PERMANENT WILTING PERCENTAGE.)

Appendix B

USEFUL TABLES AND CONVERSIONS

Table B-1. Conversion Factors for English and Metric Units

To Convert Column 1 into Column 2, Multiply by	Column 1	Column 2	To Convert Column 2 into Column 1, Multiply by
		Length	
0.621	kilometer, km	mile, mi.	1.609
1.094	meter, m	yard, yd.	0.914
0.394	centimeter, cm	inch, in.	2.54
		Area	
0.386	kilometer^2, km^2	mile^2, mi.^{2}	2.590
247.1	kilometer^2, km^2	acre, acre	0.00405
2.471	hectare, ha	acre, acre	0.405
		Volume	
0.00973	meter^3, m^3	acre-inch	102.8
3.532	hectoliter, hl	cubic foot, cu ft.^{3}	0.2832
2.838	hectoliter, hl	bushel, bu.	0.352
0.0284	liter, l	bushel, bu.	35.24
1.057	liter, l	quart (liquid), qt.	0.946
		Mass	
1.102	ton (metric)	ton (English)	0.9072
2.205	quintal, q	hundredweight cwt (short)	0.454
2.205	kilogram, kg	pound, lb.	0.454
0.035	gram, gm	ounce (avdp.), oz.	28.35
		Pressure	
14.50	bar	lb./inch^2, psi	0.06895
0.9869	bar	atmosphere, atm	1.013
0.9678	kg (weight)/cm^2	atmosphere, atm	1.033
14.22	kg (weight)/cm^2	lb./inch^2, psi	0.07031
14.70	atmosphere, atm	lb./inch^2, psi	0.06805
0.1450	kilopascal	lb./inch^2, psi	6.895
0.009869	kilopascal	atmosphere, atm	101.30
		Yield or Rate	
0.446	ton (metric)/hectare	ton (English)/acre	2.240
0.891	kg/ha	lb./acre	1.12
0.891	quintal/hectare	hundredweight/acre	1.12
1.15	hectoliter/hectare	bu./acre	0.87

Table B-2. The Metric System

The fundamental unit of the metric system is the meter (the unit of length) from which the units of mass (gram) and capacity (liter) are derived; all other units are the decimal subdivisions or multiples thereof. These three units are simply related, so that for all practical purposes the volume of one kilogram of water (one liter) is equal to one cubic decimeter.

Prefix	Meaning		Units
micro-	= one millionth,	$\frac{1}{1,000,000}$ 0.000001	
milli-	= one thousandth,	$\frac{1}{1000,}$ 0.001	
centi-	= one hundredth,	$\frac{1}{100,}$ 0.01	
deci-	= one tenth,	$\frac{1}{10,}$ 0.1	meter for length, gram for mass, liter for capacity.
unit	= one,	1.	
deka- or deca-	= ten,	$\frac{10,}{1}$ 10.	
hecto-	= one hundred,	$\frac{100,}{1}$ 100.	
kilo-	= one thousand,	$\frac{1000,}{1}$ 1000.	
mega-	= one million,	$\frac{1,000,000,}{1}$ 1,000,000.	

The metric terms are formed by combining the words "meter," "gram" and "liter" with the eight numerical prefixes.

The finer subdivisions of measurement, nano- (10^{-9}), pico- (10^{-12}) and femto- (10^{-15}) are additional prefixes sometimes usefully employed, especially with units of mass (gram) and capacity (liter).

Table B-3. Metric—U.S. System Equivalents

Length

Metric denominations and values			Equivalents in denominations in use
myriameter	=	10,000 m	= 6.2137 mi.
kilometer	=	1,000 m	= 0.62137 mi. or 3,280 ft. 10 in.
hectometer	=	100 m	= 328 ft. and 1 in.
dekameter	=	10 m	= 393.7 in.
meter	=	1 m	= 39.37 in.
decimeter	=	.1 m	= 3.937 in.
centimeter	=	.01 m	= 0.3937 in.
millimeter	=	.001 m	= 0.0394 in.

Volume

Name	No. liters	Cubic measure	Dry measure	Liquid measure
kiloliter	= 1,000	= 1 cu m	= 1.308 cu. yds.	= 264.17 gal.
hectoliter	= 100	= .1 cu m	= 2 bu. 3.35 pks.	= 26.417 gal.
decaliter	= 10	= 10 cu dm	= 9.08 qts.	= 2.6417 gal.
liter	= 1	= 1 cu dm	= 0.908 qts.	= 1.0567 qts.
deciliter	= .1	= .1 cu dm	= 6.1022 cu in.	= 0.845 gill
centiliter	= .01	= 10 cu cm	= 0.6102 cu in.	= 0.338 fluid oz.
milliliter	= .001	= 1 cu cm	= 0.061 cu in.	= 0.27 fluid dr.

Weight

Name	No. grams		Avoirdupois weight
millier or tonneau	= 1,000,000	= 1 cu m	= 2204.6 lbs.
quintal	= 100,000	= 1 hl	= 220.46 lbs.
myriagram	= 10,000	= 10 l	= 22.046 lbs.
kilogram or kilo	= 1,000	= 1 l	= 2.2046 lbs.
hectogram	= 100	= 1 dl	= 3.5274 oz.
dekagram	= 10	= 10 cu cm	= 0.3527 oz.
gram	= 1	= 1 cu cm	= 15.432 gr.
decigram	= .1	= .1 cu cm	= 1.5432 gr.
centigram	= .01	= 10 cu mm	= 0.1543 gr.
milligram	= .001	= 1 cu mm	= 0.0154 gr.

Area

hectare	=	10,000 sq. m	= 2.471 acres
are	=	100 sq. m	= 119.6 sq. yds.
centare	=	1 sq. m	= 1,550 sq. in.

Table B-4. Temperature Comparison of Celsius to Fahrenheit

Celsius (C)°	Fahrenheit (F°)
− 30	− 22
− 20	− 4
− 10	14
0	32
10	50
20	68
30	86
40	104
50	122
60	140
70	158
80	176
90	194
100	212

Conversion Formulas

$$C° = \frac{5}{9}(F° - 32) \qquad F° = (\frac{9}{5}C°) + 32$$

Table B-5. Useful Conversions

The following data are useful in calculating rates of application:

1 acre-foot of soil = 4,000,000 lbs. (approximate)
1 t. per acre = 20.8 gm per sq. ft.
1 t. per acre = 1 lb. per 21.78 sq. ft.
1 t. per acre = 25.12 quintals per hectare
1 t. per acre 6″ depth = 1 gm per 1000 gm of soil
1 gm per sq. ft. = 96 lbs. per acre
1 lb. per acre = 0.0104 gm per sq. ft.
1 lb. per acre = 1.12 kilos per hectare
100 lbs. per acre = .2296 lbs. per 100 sq. ft.
gm per sq. ft. × 96 = lbs. per acre
kg per 48 sq. ft. = t. per acre
lbs. per sq. ft. × 21.78 = t. per acre
lbs. per sq. ft. × 43560 = lbs. per acre
100 sq. ft. = 1/435.6 or .002296 acre
t. per acre-foot = 0.00136 × parts per million
cu ft. per second = 0.002228 gal. per minute
parts per million = 17.1 × gr. per gal.
parts per million × 0.00136 = t. per acre-foot

Table B-6. U.S. Weights

Troy Weight

24 grains (gr.)   = 1 pennyweight (pwt. or dwt.)
20 pennyweights = 1 ounce (oz.)
12 ounces   = 1 pound (lb.)

Apothecaries Weight

20 grains (gr.) = 1 scruple (sc.)
3 scruples   = 1 dram (dr.)
8 drams   = 1 ounce (oz.)
12 ounces   = 1 pound (lb.)

Avoirdupois Weight

27 11/32 grains (gr.)	= 1 dram (dr.)
16 drams	= 1 ounce (oz.)
16 ounces	= 1 pound (lb.)
25 pounds	= 1 quarter
4 quarters or 100 pounds (U.S.)	= 1 hundredweight (cwt.)
112 pounds (Gr. Brit.)	= 1 hundredweight
20 hundredweight or	
2000 pounds (U.S.)	= 1 ton (t.)

The U.S. short ton is 2000 pounds, the British long ton is 2240 lbs. and the metric ton (1000 kg) is 2204.6 lbs. The long ton is also used in the United States and as a measure of weight especially by steamship companies and customs officials.

Table B-7. Useful Weights and Measures for the Home Gardener and Nurseryman

Weights

Pounds per Acre	Equivalent Quantity per 100 Square Feet
100	3½ oz.
200	7½ oz.
300	11 oz.
400	14¾ oz.
500	1 lb. 2½ oz.
600	1 lb. 6 oz.
700	1 lb. 10 oz.
800	1 lb. 13 oz.
900	2 lbs. 1 oz.
1000	2 lbs. 5 oz.
2000	4 lbs. 10 oz.

Measures
(approximate)

1 level tsp.	⅙	oz.
1 level tbsp.	½	oz.
1 level c.	8	oz.
1 pt.	1	lbs.
1 qt.	2	lbs.
1 gal.	8	lbs.

Table B-8. Land Measurements

Linear Measure

1 in.	.0833 ft.
7.92 in.	1 link
12 in.	1 ft.
1 vara	33 in.
2¾ ft.	1 vara
3 ft.	1 yd.
25 links	16½ ft.
25 links	1 rd.
100 links	1 ch.
16½ ft.	1 rd.
5½ yds.	1 rd.
4 rds.	100 links
66 ft.	1 ch.
80 ch.	1 mi.
320 rds.	1 mi.
8,000 links	1 mi.
5,280 ft.	1 mi.
1,760 yds.	1 mi.
9 in.	1 span
4 in.	1 hand

Square Measure

144 sq. in.	1 sq. ft.
9 sq. ft.	1 sq. yd.
30¼ sq. yds.	1 sq. rd.
16 sq. rds.	1 sq. ch.
1 sq. rd.	272¼ sq. ft.
1 sq. ch.	4,356 sq. ft.
10 sq. chs.	1 acre
160 sq. rds.	1 acre
4,840 sq. yds.	1 acre
43,560 sq. ft.	1 acre
640 acres	1 sq. mi.
1 sq. mi.	1 section
160 acres	¼ section
36 sq. mi.	1 twp.*
6 mi. sq.	1 twp.*
1 sq. mi.	2.59 sq. km†

* twp. = township
† km = kilometer

Table B-9. Volume Measurements

Cubic Measure

1,728 cu. in.	1 cu. ft.
1 cu. ft.	7.4805 gal.
27 cu. ft.	1 cu. yd.
128 cu. ft. (4′ x 4′ x 8′)	1 cord (wood)
231 cu. in.	1 gal.
2,150.4 cu. in.	1 bu.
1.244 cu. ft.	1 bu.

Liquid Measure

1 pt. (4 gills)	16 fluid oz.
1 qt. (2 pts.)	32 fluid oz.
1 gal. (4 qts.)	128 fluid oz.
1 gal. (U.S.)	0.8327 Imperial gal.
31½ gal.	1 barrel
42 gal.	1 barrel (petroleum measure)
63 gal. (2 barrels)	1 hogshead

Dry Measure

2 pts. dry	1 qt. dry
8 qts. dry	1 pk.
4 pks.	1 bu.
105 qts. dry or 7056 cu. in.	1 standard barrel

Gallons in Square Tanks

To find the number of gallons in a square or an oblong tank, multiply the number of cubic feet that it contains by 7.4805.

Gallons in Circular Tanks

To find the number of gallons in a circular tank or well, square the diameter in feet, multiply by the depth in feet, then multiply by 5.875.

Table B-10. Convenient Conversion Factors

Multiply	By	To Get
Acres	0.4048	Hectare
Acres	43,560	Square feet
Acres	160	Square rods
Acres	4,840	Square yards
Ares	1,076.4	Square feet
Bushels	4	Pecks
Bushels	64	Pints
Bushels	32	Quarts
Centimeters	0.3937	Inches
Centimeters	0.01	Meters
Cubic centimeters	0.03382	Ounces (liquid)
Cubic feet	1,728	Cubic inches
Cubic feet	0.03704	Cubic yards
Cubic feet	7.4805	Gallons
Cubic feet	29.92	Quarts (liquid)
Cubic yards	27	Cubic feet
Cubic yards	46,656	Cubic inches
Cubic yards	202	Gallons
Feet	30.48	Centimeters
Feet	12	Inches
Feet	0.3048	Meters
Feet	0.060606	Rods
Feet	⅓ or 0.33333	Yards
Feet	0.01136	Miles per hour
Gallons	0.1337	Cubic feet
Gallons	4	Quarts (liquid)
Gallons of water	8.3453	Pounds of water
Grams	15.43	Grains
Grams	0.001	Kilograms
Grams	1,000	Milligrams
Grams	0.0353	Ounces
Grams per liter	1,000	Parts per million
Hectares	2.471	Acres
Inches	2.54	Centimeters
Inches	0.08333	Feet
Kilograms	1,000	Grams
Kilograms	2.205	Pounds
Kilograms per hectare	0.892	Pounds per acre
Kilometers	3,281	Feet
Kilometers	0.6214	Miles
Liters	1,000	Cubic centimeters
Liters	0.0353	Cubic feet
Liters	61.02	Cubic inches
Liters	0.2642	Gallons
Liters	1.057	Quarts (liquid)
Meters	100	Centimeters

(Continued)

Table B-10 (Continued)

Multiply	By	To Get
Meters	3.2181	Feet
Meters	39.37	Inches
Miles	5,280	Feet
Miles	63,360	Inches
Miles	320	Rods
Miles	1,760	Yards
Miles per hour	88	Feet per minute
Miles per hour	1.467	Feet per second
Miles per minute	60	Miles per hour
Ounces (dry)	0.0625	Pounds
Ounces (liquid)	0.0625	Pints (liquid)
Ounces (liquid)	0.03125	Quarts (liquid)
Parts per million	8.345	Pounds per million gallons water
Pecks	16	Pints (dry)
Pecks	8	Quarts (dry)
Pints (dry)	0.5	Quarts (dry)
Pints (liquid)	16	Ounces (liquid)
Pounds	453.5924	Grams
Pounds	16	Ounces
Pounds of water	0.1198	Gallons
Quarts (liquid)	0.9463	Liters
Quarts (liquid)	32	Ounces (liquid)
Quarts (liquid)	2	Pints (liquid)
Rods	16.5	Feet
Rods	5.5	Yards
Square feet	144	Square inches
Square feet	0.11111	Square yards
Square inches	0.00694	Square feet
Square miles	640	Acres
Square miles	27,878,400	Square feet
Square rods	0.00625	Acres
Square rods	272.25	Square feet
Square yards	0.0002066	Acres
Square yards	9	Square feet
Square yards	1,296	Square inches
Temperature (°C) + 17.98	1.8	Temperature, °F
Temperature (°F) − 32	5/9 or 0.5555	Temperature, °C
Ton	907.1849	Kilograms
Ton	2,000	Pounds
Ton, long	2,240	Pounds
Yards	3	Feet
Yards	36	Inches
Yards	0.9144	Meters

Table B-11. Test Weights of Agricultural Products

Commodity	Unit	Approximate Net Weight (lbs.)
Alfalfa seed	Bu.	60
Apples	Bu.	48
	Northwest box	44
	Eastern box	54
Apricots	Lug	24
	4-basket crate	24
Barley	Bu.	48
Beans, dry	Bu.	60
Beans, Lima	Bu.	56
Beans, snap	Bu.	30
Bluegrass seed	Bu.	14-30
Cherries	Lug	16
Clover seed	Bu.	60
Corn, ear (husked)	Bu.	70
Corn, shelled	Bu.	56
Cotton	Bale, gross	500
	Bale, net	480
Cottonseed	Bu.	32
Flaxseed	Bu.	56
Grain sorghums	Bu.	56
Grapefruit	Box (Fla. & Tex.)	80
	Box (Calif. & Ariz.)	65
Lemons, California	Box	79
Lentils	Bu.	60
Limes, Florida	Box	80
Meadow fescue seed	Bu.	24
Milk	Gal.	8.6
Millet	Bu.	48-50
Milo	Bu.	56
Mustard seed	Bu.	59-60
Oats	Bu.	32
Olives	Lug.	25-30
Onions, dry	Sack	50
Oranges	Box (Fla. & Tex.)	90
	Box (Calif. & Ariz.)	77
Orchardgrass seed	Bu.	14
Peaches	Lug	20
	Bu.	48
Peanuts, unshelled		
Virginia type	Bu.	22
Runners	Bu.	28
Spanish	Bu.	30
Peas, dry	Bu.	60
Potatoes	Bu.	50-60
Rice, rough	Bu.	45

(Continued)

Table B-11 (Continued)

Commodity	Unit	Approximate Net Weight (lbs.)
Rye	Bu.	56
Soybeans	Bu.	60
Sudangrass seed	Bu.	40
Sweet potatoes	Bu.	55
Timothy seed	Bu.	45
Tomatoes	Bu.	53
	Lug	32
Turpentine	Gal.	7.23
Vetch seed	Bu.	60
Walnuts	Bu.	50
Wheat	Bu.	60

Table B-12. Number of Trees or Plants on an Acre

Spacing	Number	Spacing	Number
1 by 2 ft.	21,780	6 by 8 ft.	907
1 by 3 ft.	14,520	6 by 6 ft.	1,210
1 by 4 ft.	10,890	8 by 8 ft.	680
1½ by 2 ft.	14,520	10 by 10 ft.	436
1½ by 3 ft.	9,680	12 by 12 ft.	302
2 by 3 ft.	7,260	15 by 15 ft.	194
2 by 4 ft.	5,445	16 by 16 ft.	170
3 by 4 ft.	3,630	18 by 18 ft.	134
3 by 5 ft.	2,904	20 by 20 ft.	109
3 by 6 ft.	2,420	25 by 25 ft.	70
4 by 4 ft.	2,722	30 by 30 ft.	48
4 by 6 ft.	1,815	40 by 40 ft.	27

Table B-13. Essential Growth Elements, Their Atomic Weights and Common Valence Values

Name	Symbol	Atomic Weight	Common Valence
Boron	B	10.82	3
Calcium	Ca	40.08	2
Carbon	C	12.01	4
Chlorine	Cl	35.46	-1
Copper	Cu	63.54	1, 2
Hydrogen	H	1.01	1
Iron	Fe	55.85	2, 3
Magnesium	Mg	24.31	2
Manganese	Mn	54.94	2, 4, 7
Molybdenum	Mo	95.94	3, 4, 6
Nitrogen	N	14.01	3, 5
Oxygen	O	16.00	-2
Phosphorus	P	30.98	5
Potassium	K	39.10	1
Sulfur	S	32.06	4, 6
Zinc	Zn	65.37	2

Table B-14. Atomic Weights of Elements in Common Fertilizer Materials

Name	Symbol	Atomic Weight	Name	Symbol	Atomic Weight
Aluminum	Al	26.97	Magnesium	Mg	24.31
Boron	B	10.82	Manganese	Mn	54.94
Calcium	Ca	40.08	Molybdenum	Mo	95.94
Carbon	C	12.01	Nickel	Ni	58.69
Chlorine	Cl	35.46	Nitrogen	N	14.01
Cobalt	Co	58.94	Oxygen	O	16.00
Copper	Cu	63.54	Phosphorus	P	30.98
Fluorine	F	19.00	Potassium	K	39.10
Hydrogen	H	1.01	Sodium	Na	23.00
Iodine	I	126.92	Sulfur	S	32.06
Iron	Fe	55.85	Zinc	Zn	65.37

Table B-15. Chemical Symbols, Equivalent Weights and Common Names of Ions, Salts and Chemical Amendments

Chemical Symbol or Formula	Gram Equivalent Weight	Common Name
Ca^{++}	20.04	Calcium ion
Mg^{++}	12.15	Magnesium ion
Na^+	23.00	Sodium ion
K^+	39.10	Potassium ion
Cl^-	35.46	Chloride ion
NO_3^-	62.01	Nitrate ion
NH_4^+	17.03	Ammonium ion
$SO_4^=$	48.03	Sulfate ion
$CO_3^=$	30.00	Carbonate ion
HCO_3^-	61.02	Bicarbonate ion
$CaCl_2$	55.50	Calcium chloride
$CaSO_4$	68.07	Calcium sulfate
$CaSO_4 \cdot 2H_2O$	86.09	Gypsum
$CaCO_3$	50.04	Calcium carbonate
$MgCl_2$	47.62	Magnesium chloride
$MgSO_4$	60.19	Magnesium sulfate
$MgCO_3$	42.16	Magnesium carbonate
$NaCl$	58.46	Sodium chloride
Na_2SO_4	71.03	Sodium sulfate
Na_2CO_3	53.00	Sodium carbonate
$NaHCO_3$	84.02	Sodium bicarbonate
KCl	74.56	Potassium chloride
K_2SO_4	87.13	Potassium sulfate
K_2CO_3	69.10	Potassium carbonate
$KHCO_3$	100.12	Potassium bicarbonate
S	16.03	Sulfur
SO_2	32.03	Sulfur dioxide
H_2SO_4	49.04	Sulfuric acid
$Al_2(SO_4)_3 \cdot 18H_2O$	111.08	Aluminum sulfate
$FeSO_4 \cdot 7H_2O$	139.02	Iron sulfate (ferrous)

Table B-16. Typical Composition of Manures

	Percent Moisture	Approximate Composition Pounds per Ton		
		N	P_2O_5	K_2O
Fresh manure with normal quantity bedding or litter				
Dairy	86	11	3	10
Duck	61	22	29	10
Goose	67	22	11	10
Hen	73	22	18	10
Hog	87	11	6	9
Horse	80	13	5	10
Sheep	68	20	15	8
Steer or feed yard	75	12	7	11
Turkey	74	26	14	10
Dried commercial products				
Dairy, East	10	42	63	61
Dairy, West	16	18	15	31
Hog, West	10	45	42	20
Horse	8	14	7	10
Poultry, East (with litter)	16	56	57	30
Poultry, West (droppings)	8	83	63	31
Poultry, West (with litter)	13	41	37	23
Rabbit, West	6	45	27	16
Sheep, East	10	38	30	40
Sheep, West	9	27	19	41
Stockyard, East	8	41	32	36
Stockyard, West	15	41	11	38

Table B-17. Some Acidifying Materials

Materials		Percent Sulfur Content Commercial Samples	Pounds Necessary to Equal 100 Pounds of Soil Sulfur
Soil sulfur	S	99.0	100
Sulfuric acid (98%)	H_2SO_4	32.0	3.06
Lime-sulfur solution (32° Baume)	CaS_x + water	24.0	417

Table B-18. The Approximate Amounts of Soil Sulfur (99%) Needed to Increase the Acidity of the Plow-Depth Layer of a Carbonate-Free Soil

Change in pH Desired	Pounds of Sulfur per Acre		
	Sand	Loam	Clay
8.5 to 6.5	2,000	2,500	3,000
8.0 to 6.5	1,200	1,500	2,000
7.5 to 6.5	500	800	1,000
7.0 to 6.5	100	150	300

Table B-19. Tons of Various Amendments Needed to Be Equivalent to One Ton of Sulfur

Amendment	Tons Equivalent to 1 Ton of Sulfur
Sulfur	1.00
Lime-sulfur solution, 24 percent sulfur	3.65
Sulfuric acid (98%)	3.06
Gypsum ($CaSO_4 \cdot 2H_2O$)	5.38
Iron sulfate ($FeSO_4 \cdot 7H_2O$)	8.69
Aluminum sulfate [$Al_2(SO_4)_3 \cdot 18H_2O$]	6.94

Table B-20. Plants Grouped According to Their Tolerance to Acidity

Very Sensitive to Acidity	Will Tolerate Slight Acidity	Will Tolerate Moderate Acidity	Strong Acidity Favorable
Alfalfa	Soybean	Vetch	Blueberry
Sweet clover	Red clover	Oats	Cranberry
Barley	Mammoth clover	Rye	Holly
Sugar beet	Alsike clover	Buckwheat	Rhododendron
Cabbage	White clover	Millet	Azalea
Cauliflower	Timothy	Sudan grass	
Lettuce	Kentucky bluegrass	Redtop	
Onion	Corn	Bentgrass	
Spinach	Wheat	Tobacco	
Asparagus	Pea	Potato	
Beet	Carrot	Field bean	
Parsnip	Cucumber	Parsley	
Celery	Brussels sprouts	Sweet potato	
Muskmelon	Kale	Cotton	
	Kohlrabi	Peanuts	
	Pumpkin		
	Radish		
	Squash		
	Lima, pole & snap beans		
	Sweet corn		
	Tomato		
	Turnip		
	Sorghum		

Table B-21. Amount of Limestone Needed to Change the Soil Reaction (Approximate)*

Change in pH Desired in Plow-Depth Layer	Pounds of Limestone per Acre					
	Sand	Sandy Loam	Loam	Silt Loam	Clay Loam	Muck
4.0 to 6.5	2,600	5,000	7,000	8,400	10,000	19,000
4.5 to 6.5	2,200	4,200	5,800	7,000	8,400	16,200
5.0 to 6.5	1,800	3,400	4,600	5,600	6,600	12,600
5.5 to 6.5	1,200	2,600	3,400	4,000	4,600	8,600
6.0 to 6.5	600	1,400	1,800	2,200	2,400	4,400

*A dolomitic limestone is preferable wherever there is a possible lack of magnesium.

Table B-22. The Neutralizing Value of the Pure Forms of Commonly Used Liming Materials

Material	Chemical	Neutralizing Value
		%
Calcium oxide	CaO	179
Calcium hydroxide	$Ca(OH)_2$	136
Dolomite	$CaMg(CO_3)_2$	109
Calcium carbonate	$CaCO_3$	100
Calcium silicate	$CaSiO_3$	86

Use of the neutralizing value makes possible the most simple and straight-forward comparison of one liming material with another.

Table B-23. Salt Index (Relative Effect of Fertilizer Materials on the Soil Solution)*

Material	Salt Index	Partial Salt Index per Unit of Plant Nutrient
Anhydrous ammonia	47.1	0.572
Ammonium nitrate	104.7	2.990
Ammonium nitrate–lime	61.1	2.982
Ammonium phosphate (11-48)	26.9	2.442
Ammonium sulfate	69.0	3.253
Calcium carbonate (limestone)	4.7	0.083
Calcium cyanamide	31.0	1.476
Calcium nitrate	52.5	4.409
Calcium sulfate (gypsum)	8.1	0.247
Diammonium phosphate	29.9	1.614 (N)
		0.637 (P_2O_5)
Dolomite (calcium and magnesium carbonates)	0.8	0.042
Kainit, 13.5%	105.9	8.475
Kainit, 17.5%	109.4	6.253
Manure salts, 20%	112.7	5.636
Manure salts, 30%	91.9	3.067
Monoammonium phosphate	34.2	2.453 (N)
		0.485 (P_2O_5)
Monocalcium phosphate	15.4	0.274
Nitrate of soda	100.0	6.060 (N)
Nitrogen solution, 37%	77.8	2.104
Nitrogen solution, 40%	70.4	1.724
Potassium chloride, 50%	109.4	2.189
Potassium chloride, 60%	116.3	1.936
Potassium chloride, 63%	114.3	1.812
Potassium nitrate	73.6	5.336 (N)
		1.580 (K_2O)
Potassium sulfate	46.1	0.853 (K_2O)
Sodium chloride	153.8	2.899 (Na)
Sulfate of potash–magnesia	43.2	1.971 (K_2O)
Superphosphate, 16%	7.8	0.487
Superphosphate, 20%	7.8	0.390
Superphosphate, 45%	10.1	0.224
Superphosphate, 48%	10.1	0.210
Uramon	66.4	1.579
Urea	75.4	1.618

*After L. F. Rader, Jr., et al., *Soil Sci.*, 55 : 210-218, 1943.

Table B-24.   **Tolerance of Some Plants to Chloride Levels in Saturated Extract of Soil**

Crops		Chloride*
Fruit crops	*Rootstock*	(me/l)
Citrus	⎧Rangput lime, Cleopatra mandarin	25
	⎨Rough lemon, tangelo, sour orange	15
	⎩Sweet orange, citrange	10
Stone fruit	⎧Mariana	25
	⎨Lovell, Shalil	10
	⎩Yunnan	7
Avocado	⎧West Indian	8
	⎨Mexican	5
	Varieties	
Grapes	⎧Thompson seedless, Perlette	25
	⎨Cardinal, Black Rose	10
Berries	⎧Boysenberry	10
	⎨Olallie blackberry	10
Strawberry	⎧Lassen	8
	⎨Shasta	5
Miscellaneous crops		
Beets		Near 90
Barley		90
Corn (young)		70
Aster		Near 50
Flax		50
Cotton		50
Gladiola		Near 40
Tomato		39
Geranium		Near 30
Wheat (young)		25
Gardenia		Near 25
Beans, kidney		24
Rhodesgrass		24
Sorghum		23
Alfalfa		23
Rice		23
Dallis grass		19
Beans, navy		18

*This figure is usually from four to six times greater than the chloride content in the irrigation water used.

Table B-25.   Units of Water Measurement

Volume units

One acre-inch

= 3,630 cu ft.
= 27,154 gal.
= $\frac{1}{12}$ acre-foot

One acre-foot

= 43,560 cu ft.
= 325,851 gal.
= 12 acre-inches

One cubic foot

= 1,728 cu in.
= 7.481 (approximately 7.5) gal.
weighs approximately 62.4 lbs.
(62.5 for ordinary calculations)

One gallon

= 231 cu in.
= 0.13368 cu ft.
weighs approximately 8.33 lbs.

Flow units

One cubic foot per second

= 448.83 (approximately 450) gal. per minute
= 1 acre-inch in 1 hour and 30 seconds (approximately 1 hour),
or 0.992 (approximately 1) acre-inch per hour
= 1 acre-foot in 12 hours and 6 minutes (approximately 12 hours),
or 1.984 (approximately 2) acre-feet per 24 hours

One gallon per minute

= 0.00223 (approximately $\frac{1}{450}$) cu ft. per second
= 1 acre-inch in 452.6 (approximately 450) hours, or 0.00221
acre-inch per hour
= 1 acre-foot in 226.3 days, or 0.00442 acre-foot per day
= 1 in. depth of water over 96.3 sq. ft. in 1 hour

Million gallons per day

= 1.547 cu ft. per second
= 694.4 gal. per minute

Table B-26.   Conversion Table for Units of Flow

Units	Cubic Feet per Second	Gallons per Minute	Million Gallons per Day	Acre-Inches per 24 Hours	Acre-Feet per 24 Hours
Cu ft. per second	1.0	448.8	0.646	23.80	1.984
Gal. per minute	0.00223	1.0	0.00144	0.053	0.00442
Million gal. per day	1.547	694.4	1.0	36.84	3.07
Acre-inches per 24 hours	0.042	18.86	0.0271	1.0	0.0833
Acre-feet per 24 hours	0.504	226.3	0.3259	12.0	1.0

The following approximate formulae may be conveniently used to compute the depth of water applied to a field:

$$\frac{\text{cu ft. per second x hours}}{\text{acres}} = \text{acre-inches per acre, or average depth in in.}$$

$$\frac{\text{gal. per minute x hours}}{450 \text{ x acres}} = \text{acre-inches per acre, or average depth in in.}$$

Table B-27.   Approximate Amounts of Water Held by Different Soils

Soil Texture	Inches of Water Held per Foot of Soil	Max. Rate of Irrigation Inches per Hour Bare Soil
Sand	0.5-0.7	0.75
Fine sand	0.7-0.9	.60
Loamy sand	0.7-1.1	.50
Loamy fine sand	0.8-1.2	.45
Sandy loam	0.8-1.4	.40
Loam	1.0-1.8	.35
Silt loam	1.2-1.8	.30
Clay loam	1.3-2.1	.25
Silty clay	1.4-2.5	.20
Clay	1.4-2.4	.15

Plants consume, on the average, from 0.1 to 0.3 in. of rainfall or irrigation per day.

Table B-28. Easy Conversion of Water Analysis of Ions to Pounds Material per Acre-Foot of Water

Ions	Milliequivalents per Liter (ppm)* A	Pounds per Acre-Foot of Water	Conversion Factor† B	Equivalent to Material in Pounds per Acre-Foot C	
Ca	20.0	54.4	6.8	136.0	$CaCO_3$
Mg	12.2	33.2	9.4	114.9	$MgCO_3$
Mg	12.2	33.2	13.5	164.3	$MgSO_4$
Na	23.0	62.6	6.9	159.0	NaCl
K	39.1	106.4	3.3	127.7	K_2O
K	39.1	106.4	5.2	203.2	KCl
CO_3	30.0	81.6	--	--	--
HCO_3	61.0	165.9	--	--	--
Cl	35.5	95.6	--	--	--
SO_4	48.0	130.6	0.9	43.1	S
SO_4	48.0	130.6	4.9	233.8	gypsum

* To convert me/l to lbs. per acre-foot of water, multiply by 2.72.
† A × B = C.

Table B-29. Qualitative Classification of Irrigative Waters

	Class 1 Excellent to Good	Class 2 Good to Injurious	Class 3 Injurious to Unsatisfactory
E.C.x. 10^3 at 25°C.	Less than 1.0	1.0-3.0	More than 3.0
Boron, ppm	" " 0.5	0.5-2.0	" " 2.0
Sodium percentage	" " 60	60-75	" " 75
Chloride, me/l	" " 5	5-10	" " 10

Table B-30. Permissible Limits of Boron for Several Classes of Irrigation Waters

Boron Class	Sensitive Crops	Semitolerant Crops	Tolerant Crops
	(ppm)	(ppm)	(ppm)
1	0.33	0.67	1.00
2	0.33 to 0.67	0.67 to 1.33	1.00 to 2.00
3	0.67 to 1.00	1.33 to 2.00	2.00 to 3.00
4	1.00 to 1.25	2.00 to 2.50	3.00 to 3.75
5	1.25	2.50	3.75

Boron can be leached from the soil but, if concentrations are high initially, a quantity of boron sufficient to cause trouble may remain after the concentration of other salts is reduced to a safe level.

Table B-31. Calibration of Fertilizer Application Machinery (amounts of fertilizer per 100 feet of row)

Rate per Acre	Row Width				
	18 In.	24 In.	30 In.	36 In.	48 In.
250 lbs.	14 oz.	1 lb.	1¼ lbs.	1½ lbs.	2 lbs.
500 lbs.	1¼ lbs.	2 lbs.	2½ lbs.	3½ lbs.	4½ lbs.
750 lbs.	2½ lbs.	3 lbs.	3¾ lbs.	4½ lbs.	7 lbs.
1000 lbs.	3 lbs.	4½ lbs.	5¾ lbs.	7 lbs.	9 lbs.
1500 lbs.	5 lbs.	6½ lbs.	8½ lbs.	10½ lbs.	14 lbs.
2000 lbs.	6½ lbs.	9½ lbs.	11 lbs.	13½ lbs.	18 lbs.
2500 lbs.	8½ lbs.	11½ lbs.	14½ lbs.	17 lbs.	23 lbs.
3000 lbs.	10½ lbs.	14 lbs.	17½ lbs.	21 lbs.	28 lbs.

Table B-32.   Calibration of Liquid Fertilizer Flow*

Gallons per Hour Wanted	Seconds to Fill 4-Ounce Jar	Seconds to Fill 8-Ounce Jar
½	225	450
1	112	224
2	56	112
3	38	76
4	28	56
5	22	44
6	18	36
7	16	32
8	14	28
9	12	24
10	11	22
12	9	18
14	8	16
16	7	14
18	6	12
20	5.5	11

*To calculate the rate of application required, the following formula can be used:
Acreage × lbs. or gal. per acre ÷ set time = lbs. or gal. per hour.

Table B-33.   Calibration of Fertilizer Rate Through Sprinkler Irrigation Systems

Lateral Length in Feet	No. of Sprinklers (at 40-Foot Spacing)	Area Covered by 60-Foot Setting in Acres	Quantity to Apply per Setting, for Rate of 100 Pounds per Acre
160	4	0.22	22 lbs.
240	6	0.33	33 lbs.
320	8	0.44	44 lbs.
400	10	0.55	55 lbs.
480	12	0.66	66 lbs.
560	14	0.77	77 lbs.
640	16	0.88	88 lbs.
720	18	0.99	99 lbs.
800	20	1.10	110 lbs.
880	22	1.21	121 lbs.
960	24	1.32	132 lbs.

Example

To apply fertilizer at the rate of 300 lbs. per acre with 400 ft. of lateral moved at 60-ft. setting, lbs. of fertilizer applied at each setting of the lateral are calculated as follows:

Opposite a lateral length of 400 ft., find 55 lbs. of fertilizer to be applied per setting. Multiply 55 by 3 to give 165 lbs. to apply at each setting of the lateral for 300 lbs. per acre.

Table B-34.   Estimating Amounts of Dry Bulk Fertilizer

I. Horizontal Storage

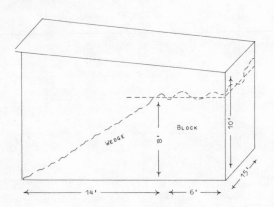

To calculate the volume in a bin.

1. Compute quantity in the block by multiplying length × width × depth.
    6′ × 15′ × 8′ = 720 cu ft.
2. Compute quantity in the wedge by multiplying length × width × depth
    and dividing by two. 14′ × 15′ × 8′ ÷ 2 = 840 cu ft.
3. Add the two volumes and multiply the total by the product density in
    lbs. per cu ft.

$$
\begin{array}{rcl}
\text{block} & = & 720 \text{ cu ft.} \\
\underline{\text{wedge}} & = & \underline{840 \text{ cu ft.}} \\
\text{total} & = & 1{,}560 \text{ cu ft.}
\end{array}
$$

1,560 × 64 (density of ammonium sulfate) = 99,840 lbs. of
product. 99,840 ÷ 2,000 = 49.92 tons.

(Continued)

Table B-34 (Continued)

II.   Cone-Shaped Pile

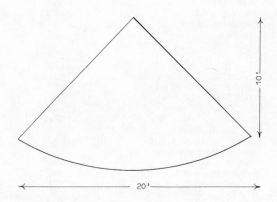

To calculate the volume in a pile:
1. Use the mathematical formula for finding the volume of a cone:

$$\frac{\pi \, r^2 h}{3}$$

Where: π is a constant = 3.14
         r is the radius (½ the diameter) of the cone measured at the base.
         h is the height of the cone

$$\frac{3.14 \times 100 \times 10}{3} = 1{,}047 \text{ cu ft.}$$

2. Multiply the volume by the product density in lbs. per cu ft.
    1,047 × 64 (density of ammonium sulfate)— = 67,008 lbs. of product—
    67,008 ÷ 2,000 = 33.5 t.

Table B-35. Product Density*

Product	Density (Lbs. per Cu. Ft.)
Ammonium nitrate, prilled	52
Ammonium nitrate, granulated	62
Ammonium sulfate	64-68
Urea	46-49
Superphosphate	68-73
Treble superphosphate	62-70
27-12-0	57
19-9-0	58
16-20-0	61-65
16-16-16	69
12-12-12	58-65
15-5-25	69
11-48-0	56-67
11-52-0	61
18-46-0	58-67
Muriate of potash	61
Muriate of potash, standard	67-75
Muriate of potash, granulated	62-68
Muriate of potash, coarse	64-69
Sulfate of potash, standard	93
Sulfate of potash, coarse	72

* The values listed are average densities which may vary slightly, depending upon compaction of the product.

INDEX

NOTES

NOTES

NOTES

NOTES

NOTES

NOTES

NOTES